BACKYARD
FARMING

Make your home a homestead

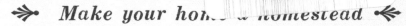

COMPOSTING

"EXPERT ADVICE MADE EASY"

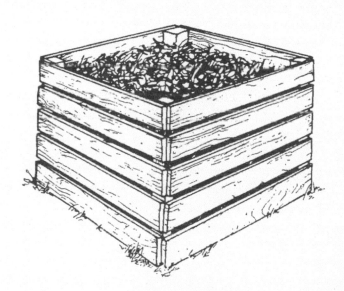

Kim Pezza

hatherleigh

Hatherleigh Press is committed to preserving and protecting the natural resources of the earth. Environmentally responsible and sustainable practices are embraced within the company's mission statement.

Visit us at www.hatherleighpress.com and register online for free offers, discounts, special events, and more.

Backyard Farming: Composting
Text copyright © 2015 Hatherleigh Press

Library of Congress Cataloging-in-Publication Data is available upon request.
ISBN: 978-1-57826-586-2

Cover Design and Interior Design by Carolyn Kasper

Printed in the United States
10 9 8 7 6 5 4 3 2 1

CONTENTS

· ·

INTRODUCTION

Whether or not you have a garden—or even a backyard—composting is an activity that anyone can do. Composting is a relatively simple activity that, while it does require some attention, is something that will literally sit quietly in the background and **peculate**, creating "black gold," which will eventually go on to restore the depleted nutrients of your soil, allowing for better plant growth within the enriched earth.

Contrary to popular belief, a good compost pile, no matter how large, does not and *should* not smell like anything except earth; it should smell like good, fresh woodlot soil. If you start to get a smell of rot or mold, something is wrong: the pile either needs to be

Photo by Kirsty Hall under the Creative Commons Attribution License 2.0.

corrected or else abandoned, and a new pile started. When done correctly, you can even do small amounts of composting indoors (red worms included)—so long as you follow all of the proper steps.

So, if you are looking for another activity that the family can work on together—especially the kids—composting certainly fits into that category. A compost pile is something that family members of all ages can work on, although you may find you need to delegate specific jobs (like turning the pile, discussed later in this book) to the bigger kids or adults, while the smaller kids will have fun adding to the pile and later sifting the completed compost, sorting out the wiggly little red worms in order to return them to the pile.

Unfortunately, some areas have banned composting, under the mistaken impression that compost heaps smell and attract rodents. But for properly executed compost, this could not be further from the truth. It is quite obvious that more education is needed to overcome these misconceptions—proper composting actually *helps* the environment, while eliminating a lot of unnecessary items going into landfills and helping to restore nutrients to the soil. You may well find yourself called upon as an advocate, teacher and soldier for the cause! You'll be surprised as to the number of people who will be thankful as you enlighten your little part of the world on the benefits of composting

This book will provide you with the basics needed to create your own compost. As with anything, there are definite do's and don'ts involved in creating your compost. There are certain conditions that your pile will prefer over others, but you shouldn't let that intimidate you; it isn't anywhere near as difficult as you might think. While it *does* take the "right balance" of certain elements to create a successful compost pile and compost, it really isn't rocket science—regardless of your personal scientific background, you can create compost!

MEET THE EXPERT

Kim Pezza grew up among orchards, muck land, dairy and beef farms, having lived most of her life in the Finger Lakes region of New York State. She has raised pigs, poultry and game birds, rabbits and goats, and is experienced in growing herbs and vegetables. In her spare time, Kim teaches workshops in a variety of areas, from art to making herb butter, oils and vinegars. She continues to learn new techniques and skills and now spends time between her grandparent's mid-1800's farm in New York and in Southwest Florida, the first and oldest cattle area in America and origin of the American cowboy.

CHAPTER 1

A BRIEF HISTORY OF COMPOSTING

· ·

Composting as we know it is the result of research, observation and discovery carried out over thousands of years. From ancient Egypt to colonial America to today, people have spent their lives trying to find ways to grow healthier, more robust crops. Our understanding of composting, and the factors that affect plant growth, come from these historical discoveries. While the historical record is somewhat sparse on the topic of composting, there are some very interesting points within the brief known history of composting, including a royal decree on composting made by a certain famous queen. So for now, let's just take a brief look at what we know about the history of composting.

The actual origins of composting are the subject of some debate. It is thought that the actual practice of composting in some form dates back to the Neolithic, Bronze and Iron Ages of Scotland, though it is not yet clear as to whether the use of compost was intentional, or an unintended consequence of general field preparation. We do know that early records, along with archeological evidence, show that both livestock and human manure were considered byproducts, and used to fertilize the soil; in fact, they seem to have been a major source of the soil's enrichment. (The arrival of sewage systems would somewhat change this, as humans began to dispose of their waste in ways other than fertilizing and composting.)

Clay tablets from the Akkadian Dynasty (2320 BCE–2120 BCE) of Mesopotamia (now Iraq), contain mention of the use of manure as fertilizer—at least 1000 years before Moses. Although it isn't known for sure, it is thought the Mesopotamians noticed plants growing much better in spots where manure had been dropped, prompting them to spread it onto the rest of their fields.

The Bible mentions the use of manure and rotted straw, as well as the same blend soaked in water (making what we would today call "compost tea"), for use in fertilizing fields. Likewise, the Talmud, a collection of laws and doctrines written by ancient Jewish teachers between 200 and 500 CE, talks about straw, ash, grass and chaff being used in the soil (along with blood from animal sacrifices) as fertilizer. Writings have been found dating back to the first century BCE demonstrating that the ancient Chinese used cooked bones as well as debris from the silk worm in their soil. There is also mention made as to the composting of straw and manure.

The Romans, Greeks and Egyptians were known to bury used straw from their animal stalls in their fields. In fact, the Roman General Marcus Porcius Cato (Cato the Elder, 234–149 BCE) discusses composting in his book *De Agri Cultura* (*Concerning the Culture of the Fields*), written around 160 BCE. In his writings, he talks about the composting of goat, sheep and cattle manure, along with straw, husks, stalks and other plant waste. It is also the first known recording of the use of worms in composting.

However, it was in ancient Egypt that composting was taken to an entirely new level, courtesy of Cleopatra. It is said that, after seeing how much worms contributed to the enrichment of soil—and what could be achieved in composting—Cleopatra issued a royal decree on the worm, declaring the worm to be sacred, protected and honored. Afraid that the removal of the worm from the Nile Valley might make the god of fertility angry, Cleopatra made the removal of the earthworm from Egypt a crime, punishable by death!

Moving forward in time, the Native Americans and European settlers in the New World used composting to ensure bountiful harvests. The Native Americans in particular made use of three methods of composting:

- **Sheet Composting**, in which the soil is layered with compostable materials
- **Composting while planting**, in which they would deposit fish or animal parts in the same hole as the seeds they were planting
- **Seed balls**, which are seeds mixed into clay along with other compostable materials. The mixture is then rolled into small balls, which are thrown to plant. The clay keeps the seed moist and protected while the compostable material provides the nutrients that the seed needs to germinate and grow.

The European settlers were known to use **fish** as fertilizer. Their "recipe" was 10 parts muck to one part fish, turning the pile (or heap) until the fish disintegrated (although the bones were found not to compost). Although this *did* work, and the result *was* an effective fertilizer, manure was more readily available to them than fish, so farmers adapted their recipe to fit with available material, revising it to two loads of muck to one barnyard of manure (or one load of muck to one stable of manure). Famous American promoters of composting include George Washington, Thomas Jefferson, James Madison and George Washington Carver.

By the mid-1800's, more "scientific" methods were being developed for composting. In 1840, German scientist Justis von Liebig proved that plants could receive nourishment from certain chemical solutions. Agriculture then began to turn from compost to artificial/chemical fertilizers.

However, in 1905 the British agronomist Sir Albert Howard, after spending nearly 30 years in India working on organic experimentation for both the farm and garden, developed what is known as the **Indore method** of composting.

Known as the modern day father of organic farming and gardening, Sir Albert discovered that the best compost had three times more green matter than manure. His method used a sandwich style of layers, building the pile up until it reached approximately five feet in height, with layers including green, manure, soil, ground limestone and rock phosphate in varying amounts. He then kept the pile moist and turned it for about three months, or until done. In 1943, his published article titled "An Agricultural Testament" renewed interest in organics and organic farming.

Other proponents of organics and composting in Europe include:

- **Rudolf Steiner**, the founder of biodynamics, which is organic farming through a holistic understanding. It is a spiritual, ethical and ecological approach not only to farming, but also to gardening, food production and nutrition.
- **Anne France-Harrar**, who created the scientific basis for compost
- **Lady Eve Balfour**, a British organic farming pioneer, educator, and proponent of composting

In the United States, J.I. Rodale was introducing Sir Albert Howard's work to American gardeners, explaining the merits of composting and soil quality. He would go on to establish a research center in Pennsylvania and create the *Organic Gardening* magazine, a wonderfully informative periodical for the organic gardener and small farmer. (Sadly, *Organic Gardening* was discontinued in 2015. If you can get your hands on past copies, it is well worth the time.)

Along with J.I. Rodale, composting was also being touted in the United States by a number of talented individuals, including E. E. Pfeiffer, a developer of biodynamic farming scientific practices, and Paul Keene, founder of Walnut Acres in Penns Creek, Pennsylvania. One of the first organic farmers in the United States, Keene produced foods sold nationally in health food stores and through mail order, including free range chickens and granola. Perhaps two of the most well-known American proponents of composting were Scott and Helen Nearing, who inspired the "back to the land" movement in the 1960s, and who continued to be influential throughout the '70s and '80s.

As you can see, while there is relatively little that we know about the exact origins of composting as a method of enriching soil, we can see that it is a practice that tends to go in and out of favor. Today, as modern consumers begin to demand fewer chemicals in their foods while pushing to recycle anything possible, composting is back on the rise, destined to become even more common on the modern backyard and urban farm.

CHAPTER 2

WHAT IS
COMPOSTING?

Composting, whether it happens in the woods behind your property or in your compost bin, is the use of natural processes to change organic materials and waste, through the process of **decomposition**, into rich soil. Anything that was once alive will decompose, and may be composted. A main ingredient in organic farming, good, composted soil will be a rich brown or black in color, coarse and crumbly in texture and rich in nutrients.

Backyard composting involves taking what happens in nature and accelerating it. While it can take nature centuries to make 1 inch of humus, composting in the backyard takes a relatively quicker few weeks to months.

Compost is created using a combination of yard debris and kitchen scraps, as well as manure when available (although

Many of the best ingredients for good compost come from the scraps your kitchen naturally produces.

manure is not necessary). Approximately 30 percent of a family's household waste may be composted. Composting reduces waste, promotes soil quality and plant productivity, and if it is created in a pile with sufficiently high temperatures, eliminates pathogens, weed seeds and deleterious (mutation) organisms, as well as sanitizing the organic waste.

When added to soil, compost conditions, fertilizes and adds humus/humic acids to the soil. Humus (not to be confused with hummus, the spread made from ground chickpeas) refers to completely decayed organic materials. When added to clay soil, it can help to alleviate drainage and compaction problems that are common when dealing with clay. In sandy soils, compost can help to improve water retention. Soil that is regularly given a boost with good compost typically needs less fertilizer than soil that is left alone, or which rarely has compost added to it.

There is, however, a difference between compost and **fertilizer**. Compost enhances and feeds the *soil*, while fertilizer feeds the *plants*. While it does add to the soil's nutrients, fertilizer meets the needs of a specific plant or plants. Basically, compost equates an individual's eating their daily meals for general nutrition, while fertilizer is like a doctor giving you a prescription for a specific need. But, compost *can* also supply small amounts of a variety of nutrients that a plant may need, such as nitrogen, phosphorus and potassium, as well as low amounts of secondary nutrients, such as calcium, sulfur and magnesium.

Compost Tea

Along with compost, you can also make something called **compost tea**. Compost tea is made from composted materials, requires no energy to make and uses relatively few materials. If you can compost (even if you don't have the space to do so) you can make compost tea.

To create compost tea, finished compost is placed in a water-permeable bag. The more compost in the bag, the stronger the compost tea. This bag will act just like a tea bag (a burlap bag works well). After tying off the top of the bag with twine, the entire bag is left in a water bucket (the amount of water is up to you) until the water turns black from the soaking compost, a process which usually takes 24–48 hours, though this may vary. Think of it like making a gigantic cup of "tea," but for the garden. This "tea" contains beneficial organisms, microbials and micronutrients, and as a result may help in suppressing disease in your garden.

The compost tea should be applied to your plants undiluted. It may be sprayed on or poured using a watering can, and will not burn leaves or roots. Apply as often as you like, whenever you like (as long as the ground is not frozen, as this will prevent it from penetrating). The spent compost may be removed from the "tea bag" and mixed into the garden soil.

Whether you use compost or compost tea, composting organic materials/waste allows you to not only return needed organisms to the soil, but provide additional nutrients as well. Not to mention, you help to keep waste out of landfills—that 30 percent of repurposed household waste really adds up!

CHAPTER 3

SELECTING AND SETTING UP A SYSTEM

. . . .

Now that you've made the decision to set up a composting system, there are a number of options open to you, including both commercial and homemade systems. In making your decision, the first thing you'll need to consider is the type of space you have available for use. It doesn't matter if you have 30 acres or a studio apartment—you can compost. However, the amount of space that you have available *will* dictate what options are available.

If you are fortunate enough to have a decent-sized backyard, your composting options are varied, and you can go with whatever method works best for you. You can choose to go the homemade route, which can be as simple as burying your compost ingredients in your backyard; or, you can choose to pursue the commercial route, purchasing pre-made, artificial composting systems.

First, let's look at some of the homemade composting systems—specifically, soil incorporation, grasscycling, covered wind rows and compost bins/enclosures.

Soil Incorporation

While composting through **soil incorporation** (also known as **trench composting**) basically just involves burying your organic materials (usually food scraps), there *is* a bit of strategy involved. This method is best suited for smaller amounts of composting and, when done properly, does not attract pests.

Compostable materials suitable for this method include:
- grains and cereals
- fruits and vegetables
- tea bags
- coffee grounds

Items that should *not* be composted include:
- dairy
- oil and grease
- meat and bone

To incorporate, you'll first need to dig a trench. Make it as long and as wide as you like, but at least 12–15 inches deep. Note that 12–15 inches results in plenty of open trench space, so make sure to take account of how much material you plan to compost. Don't make extra work for yourself; match your trench to your composting needs. If 12–15 inches isn't enough (though it should be plenty deep) you'll need to plan accordingly, either with a deeper or longer trench.

Do not discard the soil that you have removed from the trench, as you will need to fill your trench back in. You may dig these trenches around trees and shrubs (being mindful not to disturb any root systems) or gardens. This will allow them to take advantage of the nutrients your compost will create. You can also locate trenches in any areas that you plan to turn into a small garden, plant with berry bushes, etc. Then, when you plant over the trench later on (once the composting cycle has completed) the new plantings can make use of the newly composted soil.

Before placing your compostable material in the trench, chop the food/organic waste and mix it with a little of the soil that you removed from the trench. This will help to hasten the decomposition process. Place this mix in the bottom of the trench, and then cover with the rest of the soil. Note that you should place at least 8 inches of soil over this mix when burying it, to prevent any animals from being attracted to the scent and digging the trench up. This will also help to prevent any smell from permeating the air.

Breakdown/composting time in a soil incorporation system can take anywhere from one month to one year, depending on conditions, microorganisms in the soil, temperature and the soil itself.

Grasscycling

Simply put, **grasscycling** is the act of leaving grass clippings on the lawn (or else using them as mulch), allowing the grass to decompose and act as a fertilizer supplement.

The term "grasscycling" first came into use in the 1990s, when efforts began to try to reduce the amount of grass clippings going into the landfill each year. However, simply adding all of one's grass clippings into a home compost pile/bin/system can easily overwhelm the system. In a single year, a 1000 square foot lawn can produce at least 200 pounds of clippings.

This is where grasscycling comes in.

Grass is approximately 80 percent water and, with its high nitrogen content, breaks down easily releasing the nitrogen (along with other nutrients) back into the lawn. This process usually completes itself within 1–2 weeks.

The typical fertilizer count of lawn clippings yields:

- Nitrogen: 4 percent
- Potassium: 2 percent
- Phosphorus: 0.5 percent

Grasscycling can thereby provide a lawn with up to 50 percent of the fertilizer that it needs, including up to 20 percent (or more) of its yearly nitrogen requirements.

Some tips for your grasscycling project include:

- Cutting the grass when it is between 3–4 inches high, without cutting more than a third of its length
- Mow the lawn when it is dry to the touch
- Make sure the lawnmower blades are sharp
- If you can, use a mulching mower; although not absolutely necessary, a mulching mower chops clippings into small pieces, helping the decomposition process to occur a bit faster

No matter the size of your lawn, grass clippings are a viable option. Not only will you feed your lawn, you'll be keeping clippings out of the landfills, along with the plastic lawn bags used to store them.

Covered Wind Row

Covered wind rows are long, heaped compost piles at least 4 feet in height (although they can be higher), with an ideal width of 14–15 feet. The pile needs to be high enough to generate sufficient heat and maintain its interior temperature for the composting process, but also not so large that it will prevent the flow of oxygen to the core of the pile. The waste is applied in layers, and then covered with either plastic or a membrane. Larger facilities may keep their wind rows under a roof.

Although typically an agricultural process best suited for larger volumes of composting (such as one might find on a larger farm) smaller farms and homesteads with at least some acreage available may be able to take advantage of a miniaturized version of this type of composting.

While not ideally suited to backyard composting, some larger scale rural backyard farms may be able to benefit from this method. Photo by themaria under the Creative Commons Attribution License 2.0.

The rows are then turned (with rows of this size, turning is usually done mechanically) to improve oxygen content, as well as to redistribute any hot or cold spots within the pile, mixing in or removing any moisture. Wind row composting can use organic matter, including food scraps (if buried deep enough in the pile), manure and crop residue (if you are farming).

Materials suitable for wind row composting include:
- coffee grounds
- fruits and vegetables
- grains and cereals
- tea bags

Items that should *not* be composted include:
- dairy
- meat and bone
- oil and grease

Although covered wind row composting is definitely not for the urban or suburban backyard, if you do have some space, you might want to play around a bit with the wind row method—on a smaller scale. As always, find what will work right for you in terms of size and content, with size being relevant to the space you have available and the percentage you personally wish to use; there is no "right" size.

Composting Bins/Enclosures

There are different styles of composting bins, though they typically fall under the categories of either commercial or homemade/DIY. Regardless of category, the styles include:

- Stationary
- Tumblers
- Worm composters (also known as vermicomposting)
- Indoor composters

Stationary bins are usually the largest composting bins, and therefore hold the most composting materials. They are usually built on and remain in one spot.

Stationary bins, like this one, are ideally suited for processing larger amounts of material, and producing larger quantities of compost.

Tumblers are barrel-style composters that are designed to make the turning process easier, as they allow the gardener to spin or roll the barrels. However, they do not hold as much as a stationary compost bin can.

Worm composters are, much as the name suggests, composting systems that use worms to perform the decomposition process (specifically, a particular variety of red worm)—a process known as vermicomposting. Although any good outdoor compost pile has the chance to attract red worms, worm composting relies on the worms doing the bulk of the work. Many argue that vermicomposting is the best composting system; for this reason, we'll be discussing this method in more detail in the next chapter.

Indoor composters are, simply put, composting systems that you can keep indoors. These are especially good for those who want to compost but have little to no outdoor space. Note that some indoor composters can also be worm compost systems.

Indoor composters are available (or can be made) to fit even small spaces, like condos and apartments. Just because you don't have a yard doesn't mean you can't compost! Photo by Mathias Baert under the Creative Commons Attribution License 2.0.

Although there are a number of options available when looking for composting bins, with each style coming in a variety of shapes, sizes and price ranges, you will likely find that one system works better for your needs or circumstances. Do your homework; talk with neighbors, friends or family who compost and, if you are looking to purchase a commercial composter, go to the stores whenever possible and actually take a look at the bins.

Homemade Composting Bins

Should you have the time, energy and resources to make your own bin, you will often find that not only can you repurpose and reuse existing materials for your bin, you can usually save a substantial bit of money by doing so. Most DIY designs only take a day to construct, and require only one person

Don't be afraid to get creative! The sky's the limit when it comes to composting bins—feel free to take an existing design and really make it your own. Photo by Andy Wright under the Creative Commons Attribution License 2.0.

to put it together (although having a helper will make the job easier and faster).

If you want to compost using a bin in your backyard, first decide whether you want to try fashioning a composting bin from either new or recycled items, and whether you'd like to use someone else's directions, or create one of your own design. Any of the compost bin styles mentioned previously can be built as a DIY project, most in a few hours

Even the most basic set-up, so long as it holds the compost and keeps it away from pests, can be suitable for use in composting. Photo by Bryan Alexander under the Creative Commons Attribution License 2.0.

to a day. Note that composting bins are not necessary to get good compost, as you can accomplish the same results by just making a plain old pile or heap. The advantage of bins is that they help to keep the compost materials together and allow for heat to build up inside the pile.

Compost bins may be of any size, and may be made from recycled materials such as old pallets and wood. However, plywood should *not* be used in construction of compost bins, as the moisture that your compost expels will de-laminate the plywood.

Your composting bin doesn't need to be anything fancy—it just needs to work, and work for you! Photo by Cuifen Peu under the Creative Commons Attribution License 2.0.

Make sure that any lumber you choose has not been treated, as toxins from the treatment can seep into your compost. When using newly purchased wood, you can just check labels; however, if you are reusing/re-purposing old wood, you will need to be more vigilant. Bins can also be made from plastic tubs and plastic barrels, brick, garbage cans, and just about anything else you can think of, provided it can hold to your chosen design.

Homemade Stationary Bins

If you have the space, then you will most likely want to consider a **stationary bin.** These bins, especially if you build them in larger sizes, will most likely be a DIY project. Stationary bins are basically a box, usually (but not always) made of wood. They should be large enough to allow for turning of the composting material using a shovel or pitch fork, which will aerate the pile and speed up the composting process. These bins may be made of wire, brick, concrete block, or anything else that will hold its shape (within limits; see above for materials to avoid when constructing composting bins).

Many people will build two or three bins connected together; if you have a double bin, when it comes time to turn your compost you can turn it right into the empty bin next door.

A triple bin will even allow you to turn an old pile into the second bin, start a new compost pile in the first bin, and still have a spare bin for turning. Although not strictly necessary, stationary bins should also have covers to avoid over-soaking the compost from rain (or snow); if the material gets too wet, the extra moisture will slow down the composting process.

Considering the sensitivity of compost piles to too much moisture, a cover is necessary for any stationary bin out in the elements. Photo by Solylunafamilia under the Creative Commons Attribution License 2.0.

Wooden Box Bin

A wooden box bin can be built inexpensively using wooden pallets. Or you can use lumber to make a nicer looking bin. The costs will vary, depending on whether you use pallets or new lumber. Used pallets are often available from manufacturers and landfills.

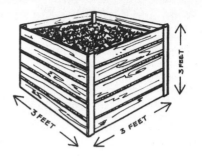

What You Need

Materials

- 4 wooden pallets (5 pallets if you want a bottom in the container), sized to make a four-sided container at least 3 feet x 3 feet x 3 feet
- nails
- wire ties

or

- 1 12-foot length of 2 x 4 lumber
- 5 12-foot lengths of lumber, 6 x ¾
- nails

Tools

- saw
- sledge hammer
- work gloves

Building a Wooden Box Bin

If using wooden pallets:

1. Nail or wire four pallets together to make a four-sided container at least 3 feet x 3 feet x 3 feet. The container is ready to use.
2. A fifth pallet can be used as a base to allow more air to get into the pile and to increase the stability of the bin.

If using lumber:

1. Saw the 12-foot length of 2 x 4 lumber into four pieces, each 3 feet long, to be used as corner posts.
2. Choose a 3-foot-square site for your compost bin, and pound the four post3 into the ground 3 feet apart, at the corners of the square.
3. Saw each of the five 12-foot boards into four 3-foot pieces. Allowing five boards to a side and starting at the bottom, nail the boards to the posts to make a four-sided container. Leave ½ inch between the boards to allow air to get into the pile.
4. If you wish to decrease your composting time, build a second holding unit so the wastes in one can mature while you add wastes to the other.

Adding Wastes

Add wastes as they become available. Nonwood materials such as grass clippings and garden weeds work best. You can speed up the process by chopping or shredding the wastes. If you have two units, when the first unit is full let the compost mature and add wastes to the second unit.

Maintaining Your Compost Pile

Although you do not need to turn this pile, make sure that it is moist during dry spells. Compost should be ready in about one year.

From *COMPOSTING: Wastes to Resources*, Cornell Cooperative Extension

Wood and Wire Three-Bin Turning Unit

A wood and wire three-bin turning unit can be used to compost large amounts of yard, garden, and kitchen wastes in a short time. Although relatively expensive to build, it is sturdy, attractive and should last a long time. Construction requires basic carpentry skills and tools.

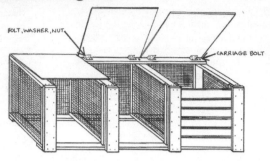

What You Need

Materials

- 4 12-foot (or 8 6-foot) lengths of pressure-treated 2 x 4 lumber
- 2 10-foot lengths of pressure-treated 2 x 4 lumber
- 1 10-foot length of construction grade 2 x 4 lumber
- 1 16-foot length of 2 x 6 lumber
- 6 8-foot lengths of 1 x 6 lumber
- 1 4-x-8-foot sheet of ½-inch exterior plywood
- 1 4-x-4-foot sheet of ½-inch exterior plywood
- 22 feet of 36-inch-wide ½-inch hardware cloth
- 2 pounds of 16d galvanized nails
- 250 poultry wire staples (or a power stapler with 1-inch galvanized staples)
- 12 ½-inch carriage bolts 4 inches long
- 12 washers and 12 nuts for the bolts
- 6 3-inch zinc-plated hinges
- 24 washers and 24 nuts for the hinges
- 1 quart wood preservative or stain

Tools

- tape measure
- hand saw or circular power saw
- hammer
- tin snips
- carpenter's square
- optional: power stapler with 1-inch galvanized staples
- drill with ½-inch bit
- screwdriver
- ¾-inch socket or open-ended wrench
- pencil
- safety glasses
- ear protection
- dust mask
- work gloves

Building a Wood and Wire Three-Bin System

1. Cut two 31 ½-inch and two 36-inch pieces from a 12-foot length of pressure-treated 2 x 4 lumber. Butt joint and nail the four pieces into a 35-inch x 36-inch "square." Repeat, building three more frames with the remaining 12-foot lengths of 2 x 4 lumber.

2. Cut four 37-inch lengths of hardware cloth. Fold back the edges of the wire 1 inch. Stretch the pieces of hardware cloth across each frame. Make sure the corners of each frame are square and then staple the screen tightly into place every 4 inches around the edge. The wood and wire frames will be dividers in your composter.

3. Set two dividers on end 9 feet apart and parallel to one another. Position the other

two dividers so they are parallel to and evenly spaced between the end dividers. The 36-inch edges should be on the ground. Measure the position of the centers of the two inside dividers along each 9-foot edge.

4. Cut a 9-foot piece from each 10-foot length of pressure-treated 2 x 4 lumber. Place the two treated boards across the tops of the dividers so each is flush against the outer edges. Measure and mark on the 9-foot boards the center of each inside divider.

5. Line up the marks, and through each junction of board and divider, drill a ½-inch hole centered 1 inch in from the edge. Secure the boards with carriage bolts, but do not tighten them yet. Turn the unit so the treated boards are on the bottom.

6. Cut one 9-foot piece from the 10-foot length of construction grade 2 x 4 lumber. Attach the board to the back of the top by repeating the process used to attach the base boards. Using the carpenter's square or measuring between opposing corners, make sure the bin is square. Tighten all the bolts securely.

7. Fasten a 9-foot length of hardware cloth to the back side of the bin with staples every 4 inches around the frame.

8. Cut four 36-inch-long pieces from the 16-foot length of 2 x 6 lumber for front runners (Save the remaining 4-foot length.) Rip cut two of these boards to two 4¾-inch wide strips. (Save the two remaining strips.)

9. Nail the 4¾-inch-wide strips to the front of the outside dividers and baseboard so they are flush on the top and the outside edges. Center the two remaining 6-inchwide boards on the front of the inside dividers flush with the top edge and nail securely.

10. Cut the remaining 4-foot length of 2 x 6 lumber into a 34-inch-fong piece and then rip cut this piece into tour equal strips. Trim the two strips saved from step 8 to 34 inches. Nail each 34-inch strip to the insides of the dividers so they are parallel to and 1 inch away from the boards attached to the front. This creates a 1-inch vertical slot on the inside of each divider.

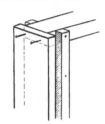

11. Cut the 6 8-foot lengths of 1 x 6 lumber into 18 slats, each 31¼ inches long. Insert the horizontal slats, 6 per bin, between the dividers into the vertical slots.

12. Cut the 4-x-8-foot sheet of exterior plywood into two 3-x-3-foot pieces. Cut the 4-x-4-foot sheet of exterior plywood into one 3-x-3-foot piece. Center each 3-x-3-foot piece on one of the three bins and attach each to the back top board with two hinges.

13. Stain all untreated wood.

Adding Wastes

Do not add wastes as they become available with this system. Collect enough wastes to fill one of the three bins at one time. You can collect woody as well as nonwood wastes. Add thin layers of different kinds of organic materials or mix the wastes together.

Before adding new wastes to an empty bin, collect enough to fill the entire bin.

Maintaining Your Compost Pile

Take the temperature of your pile every day. After a few days, the temperature should reach between 130° and 140°F (54° to 60°C). If your pile gets very hot, turn it before the temperature gets above 155°F (68°C). In a few days, the temperature will start to drop. When the temperature starts going down, turn your compost pile into the next bin with a pitchfork. The temperature of your compost pile will increase again and then, in about four to seven days, start to drop. Turn your compost pile into the third bin. The total time for composting should be less than one month.

From *COMPOSTING: Wastes to Resources*, Cornell Cooperative Extension

Cinder Block Bin

A cinder block bin is sturdy, durable, and easily accessible. If you have to buy the cinder blocks, it is slightly more expensive to build than the wire mesh or snow fence bins.

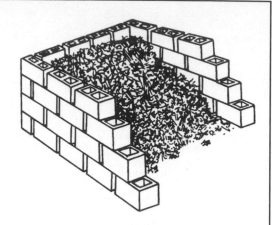

What You Need

- about 46 cinder blocks for the first bin
- optional: about 32 blocks for a second bin
- work gloves

Building a Cinder Block Bin

1. Place 5 cinder blocks in a row along the ground at your composting site, leaving about ½ inch between each block to let in air.
2. Place 4 cinder blocks in another row along the ground perpendicular to and at one end of the first row, forming a square corner; leave about ½ inch between each block.
3. In the same way, place 4 cinder blocks at the opposite end of the first row to form a three-sided enclosure.
4. Add a second layer of blocks, staggering them to increase stability and leaving about ½ inch between each block. There should be a layer of 4 cinder blocks on each of the three walls of the enclosure.
5. Add a third layer of blocks, again staggering them to increase stability, with 5 blocks across the back of the enclosure and 3 on each side.
6. The last and top layer should have 4 blocks across the back and 3 on each side.
7. Optional: If you wish to decrease your composting time, build a second bin next to the first so the wastes in one can mature while you add wastes to the other. Use one side wall of the first bin so you only need to build two additional walls.

Adding Wastes

Add wastes as they become available. Nonwood materials such as grass clippings and garden weeds work best. You can speed up the process by chopping or shredding the wastes. If you have two units, when the first unit is full let the compost mature and add wastes to the second unit.

Maintaining Your Compost Pile

Although you do not need to turn this pile, make sure that it is moist during dry spells. Compost should be ready in about one year or more.

From *COMPOSTING: Wastes to Resources*, Cornell Cooperative Extension

Cinder Block Turning Unit

A cinder block turning unit looks like three cinder block holding units in a row. It is sturdy, and if you can find used cinder blocks, it is inexpensive to build.

What You Need

• about 98 cinder blocks
• work gloves

Building a Cinder Block Turning Unit

1. Place 12 cinder blocks in a row along the ground at your composting site, leaving about ½ inch between each block to let in air.
2. Place 4 cinder blocks in another row along the ground perpendicular to and at one end of the first row, forming a square corner: leave about ½ inch between each block.
3. In the same way, place 4 cinder blocks at the opposite end of the first row to form a three-sided enclosure.
4. Place two more rows—4 cinder blocks each—along the ground, parallel to the ends and evenly spaced within the enclosure. This divides the enclosure into three separate bins.
5. Add a second layer of blocks, staggering them to increase stability and leaving about ½ inch between each block. There should be a layer of 13 cinder blocks across the back and 3 cinder blocks on the sides of each bin.
6. Add a third layer of blocks, again staggering them to increase stability with 12 blocks across the back of the enclosure and 3 on each side.
7. The last and top layer should have 13 blocks across the back and 2 on each side.

Adding Wastes

Do not add wastes as they become available with this system. Collect enough wastes to fill one of the three bins at one time. You can collect woody as well as nonwood wastes. Add thin layers of different kinds of organic materials or mix the wastes together.

Before adding new wastes to an empty bin, collect enough to fill the entire bin.

Maintaining Your Compost Pile

Take the temperature of your pile every day. After a few days, the temperature should reach between 130° and 140°F (54° to 60°C). If your pile gets very hot, turn it before the temperature gets above 155°F (68° C). In a few days the temperature will start to drop. When the temperature starts going down, turn your compost pile into the next bin with a pitchfork. The temperature of your compost pile will increase again and then, in about four to seven days, start to drop. Turn your compost pile into the third bin. Continue to take the temperature and turn the compost pile until the compost is ready. The compost should be ready in about one or two months.

From *COMPOSTING: Wastes to Resources*, Cornell Cooperative Extension

Wire Mesh Bin

A wire mesh bin is inexpensive and easy to build out of either galvanized chicken wire or hardware cloth. (Nongalvanized chicken wire can also be used but will not last very long.) Posts provide more stability for a chicken wire bin, but make the bin difficult to move. A wire mesh bin made without posts is easy to lift, allowing you to get at the compost that is already "done" at the bottom of the pile while the top of the pile is still cooking.

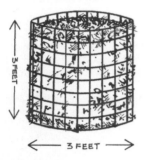

What You Need

Materials
- 12½ feet of 36-inch-wide 1-inch galvanized chicken wire or ½- inch hardware cloth
- heavy wire for ties
- 3 or 4 4-foot wooden or metal posts (for chicken wire bin)

Tools
- heavy-duty wire or tin snips
- pliers
- hammer (for chicken wire bin)
- metal file (for hardware cloth bin)
- work gloves

Building a Wire Mesh Bin

If using chicken wire:
1. Fold back 3 to 4 inches of wire at each end of the cut piece to provide a strong, clean edge that won't poke or snag and which will be easy to latch.
2. Stand the wire in a circle and set it in place for the compost pile.
3. Cut the heavy wire into lengths for ties. Attach the ends of the chicken wire together with the wire ties, using pliers.
4. Space wood or metal posts around the inside of the chicken wire circle. Holding the posts tightly against the wire, pound them firmly into the ground to provide support.

If using hardware cloth:
1. Trim the ends of the hardware cloth so the wires are flush with a cross wire to get rid of edges that could poke or scratch hands. Lightly file each wire along the cut edge to ensure safe handling when opening and closing the bin.
2. Bend the hardware cloth into a circle, and stand it in place for the compost pile.
3. Cut the heavy wire into lengths for ties. Attach the ends of the hardware cloth together with the wire ties, using pliers.

Adding Wastes

Add wastes as they become available. Nonwood materials such as grass clippings and garden weeds work best. You can speed up the process by chopping or shredding the wastes.

Maintaining Your Compost Pile

As you keep adding wastes to the wire mesh bin, the material at the bottom will become compost sooner than the material at the top. If you want to use the compost at the bottom of the pile, you can remove the wire holding unit and place it next to the pile. Then, use a pitchfork to move the compost back into the moved holding unit, adding the material from the top of the pile first. Continue until you have replaced all the compost. Now the compost at the top of the bin is ready to use.

You also can scoop finished compost from the bottom of the pile by lifting one side of the unit.

Although you do not need to turn this pile, make sure it is moist during dry spells. Compost should be finished in about one year.

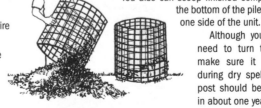

From COMPOSTING: Wastes to Resources, Cornell Cooperative Extension

Snow Fence Bin

A snow fence bin is simple to make. It works best with four posts pounded into the ground for support.

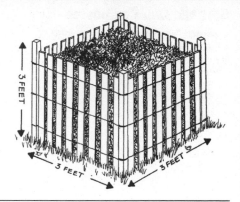

3 FEET

3 FEET

3 FEET

What You Need

Materials
- 4 wooden or metal posts, at least as tall as the snow fence
- heavy wire for ties
- 12½ feet of snow fencing, at least 3 feet tall

Tools
- heavy-duty wire or tin snips
- pliers
- sledge hammer
- work gloves

Building a Snow Fence Bin

1. Choose a 3-foot-square site for your compost bin, and pound the four wooden or metal posts into the ground 3 feet apart at the corners of the square.
2. Cut the heavy wire into lengths for ties. Attach the snow fence to the outside of the posts with the wire ties, using pliers.
3. Attach the ends of the snow fence together in the same way, forming a 3-foot-square enclosure.

Adding Wastes

Add wastes as they become available. Nonwood materials such as grass clippings and garden weeds work best. You can speed up the process by chopping or shredding the wastes.

Maintaining Your Compost Pile

Although you do not need to turn this pile, make sure that it is moist during dry spells. Compost should be ready in about one year. Simply remove the fencing and the compost is ready to use.

From *COMPOSTING: Wastes to Resources*, Cornell Cooperative Extension

Garbage Can Composter

A garbage can composter is inexpensive and easy to build. It can be used for food or garden wastes. You do, however, need to turn the wastes.

What You Need

Materials
- garbage can with cover
- coarse sawdust, straw, or wood chips

Tools
- drill
- pitch fork, shovel, or compost turner
- work gloves

Building a Garbage Can Composter

1. Drill three rows of holes 4 to 6 inches apart all around the sides of the garbage can. Then drill several holes in the base of the can. The holes allow air movement and the drainage of excess moisture.
2. Place 2 to 3 inches of dry sawdust, straw, or wood chips in the bottom of the can to absorb excess moisture and let the compost drain.

Adding Wastes

Add fruit, vegetable, and garden wastes. Make sure not to add too much of any one waste at a time.

Maintaining Your Compost Pile

Regularly mix or turn the compost with a pitch fork, shovel, or compost turner and keep it covered. This adds air and mixes L1P the different wastes, preventing the compost from getting smelly. A smelly compost pile may attract animals and cause neighbors to complain.

From *COMPOSTING: Wastes to Resources*, Cornell Cooperative Extension

Indoor/Outdoor Container Bins for Small Spaces

If you don't have lots of room—say, if you just have a porch or balcony—you can still make compost bins from covered plastic storage containers. These containers should be around 18 gallons or larger in size, new or used. (If they are used, make sure nothing toxic was stored in them. This usually isn't a problem with covered containers; however it doesn't hurt to check before you use them.) You can use any color you want, but as a general rule the darker it is, the better it is for heat and moisture retention. Do *not* use clear containers; if all you can get are clear, then spray paint the outsides in a dark color, on the outside.

You can put together a kitchen compost bin with relatively little effort, and common materials— some of which you may already have lying around, left over from other projects. Photo by Mike Luberman under the Creative Commons Attribution License 2.0.

Next, drill holes in the container, approximately 2 inches apart. The holes may be any size, but if you cut them too large, you will need to line the container with a rustproof screening to prevent any insects or small animals from invading the bin. You also want to make sure that the compost doesn't come spilling out! The holes should be on the sides, lid and on the bottom. There are some more complicated bins that you can make, and you can find links to more in the Resources section of this book, but as a general guide, that's really all it takes.

Once your plastic composting bin is ready to fill, be sure to chop up the material you plan to put in into small pieces. This allows for a faster breakdown process. In terms of upkeep, make sure to shake the bin daily. If you find that your compost bin has an odor or is too wet, add some sawdust (making sure it isn't from treated wood) or shredded newspaper. If, on the other hand, you find that your mix is too dry, then either wet it down with water or add additional, moisture-rich ingredients. When your compost is done, clean it through with a sieve before using it to remove chunks of un-composted matter (an old window screen works great!)

Homemade Tumblers

Some prefer to make tumbling compost bins. There are a number of ways to create these, and they start off similar to creating plastic composting bins. First, you will need to find your container—in this case, a barrel (one that has not been used for anything toxic). The barrel may be plastic or metal, although plastic will be a bit lighter (remember that you

These tumblers, which are based on homemade designs, demonstrate the ease with which you can mix your compost using a tumbler. Photo by Bev Wagar under the Creative Commons Attribution License 2.0.

will be physically moving your tumbling compost bin to mix your organic matter). But if all you can find is a metal barrel, it will still work just fine.

With the barrel on its side, cut a door into it. This is done by cutting a rectangle out of the side of the barrel, then reattaching it on one long side with a couple hinges. Then, attach a fastener that will keep the door closed tight, making sure that organic matter/compost won't come spilling out when tumbling. This door may be any size that you want, but keep in mind that you will be shoveling compost out of the barrel, so unless you want to remove the matter with a small scoop, try to make the door large enough for at least a small shovel.

Next, drill holes in the barrel. Some will just drill holes around the barrel, leaving the door and ends alone; others will drill holes in these as well. It's up to your personal preference.

From this point on, there are a number of variations in tumbler design. Some will just leave the barrel as it is and roll it around the grass to turn the material. Some take this one step further, and fit each end of the barrel with a tire to make it easier to roll (and to keep the barrel off the ground). The barrel can be kept in place and prevented from rolling the rest of the time by using a wedge, brick or anything else that will act as a stop. Others will cut holes in the middle of each end, run a pipe through the barrel, and then set it into "x" shaped brackets, on each end, allowing for easier turning.

There are many examples of tumbler plans, both online as well as in composting books and homesteading/gardening publications. There are also some links in the Resource section of this book.

Commercial Composting Bins

Although homemade bins will allow you make specifically what you want or need (as well as keeping to your price range by using materials that you have on hand or can otherwise scrounge up), some gardeners or farmers may not want to build their own bins, or else may not have the time to build them. Fortunately, as composting has become more mainstream, more and more farm stores, local greenhouses, garden stores, and even some of the "big box" stores now carry compost bins (there are also many places online that sell composting bins).

There are dozens of brands of composters in all shapes, sizes, styles, materials and colors available commercially, with most being constructed from plastic of some form, or else wood. You can find these commercially made bins for the outdoors, indoors, and even indoor/outdoor hybrids, as well as worm-powered compost bins. Most manufacturers will have their own small variations or selling points unique to their system, so feel free to shop around and find the brands or models that appeal to you and suit your needs.

Just make certain to do your research before making any final purchases, including checking to see if there are any common issues or hidden costs involved in installing or maintaining your selected composter.

Finding the Right Location

After you have selected the compost bin system that you want—whether it is a homemade system or a store-bought commercial system—you will need to figure out where you want to keep it (if you haven't already). Should you decide on a stationary bin system, make sure you select the correct place the first time. A stationary system is just that—stationary; it stays in one place. Of course, you *can* build it such that the bins can be moved at some point; however, should you decide that you want—or need—to move the system, you will either need to wait for the bins to be empty, or else remove the organic matter and place in a few wheelbarrows or buckets, after which you can tear the bins apart, relocate and rebuild them, and finally refill them with the organic matter that you removed. In other words, it's much easier to simply place the system in the right place the first time. The same goes for freestanding compost piles and trenches.

When placing your outdoor compost system, both sunny and shaded locations are viable; however, composting in full sun will hasten the composting process. Note that if you are using a system dependent on vermicomposting (discussed in the next chapter) your pile should be in the shade, as your worms will not be able to deal with the heat. If you have limited space, you will obviously have fewer choices as to the positioning of the system.

However, if you do have plenty of space and have multiple areas where your bins will work, keep this in mind—you *will* need to add to and tend the bins. Although it may seem like a great idea to have the bins as far away as possible, don't forget the old adage—out of sight, out of mind. If the bins are too far away from the house, you may get a bit lazy about tending the bins, or letting the scraps build up for too long in the kitchen before you take them out.

As you may need to occasionally add water to your compost bin, try to locate the bins near a water source. You don't need to be right next to the faucet, but you don't want to be hauling water halfway across the farm either. In other words, take everything into consideration during placement, and look to find a happy medium. Remember that proper compost should not smell any different than woodlot soil, so even if the bin is near the house, you still will be able to open your windows and spend time in the yard.

If you plan on placing your compost bin in the house, first make sure that it is built for indoor use, whether homemade or commercially built. There are small composting buckets and certain vermicomposting systems that are made specifically for indoor use. If you are making your own indoor system, make sure that any DIY plans you follow are appropriate for the indoors. You want your organic materials properly contained, and you certainly don't want liquids leaking out all over your floor. You'll also want to make sure that the system fits in your space. You don't want your friends and family tripping over your compost bin!

When setting up your commercial bins, make sure to follow the manufacturer's directions for putting them together. If you run into difficulties, you may be able to adapt some systems to work for you, but there's no guarantee your modifications will work. The advantage of building your own system is that you are far better able to make tweaks and changes as needed.

In the end, when selecting and setting up a system, you will have at least a few choices available to you, especially if you have a good deal of space to work with. Even if you are restricted in terms of space, there are still options in regards to the types of systems available. If you're feeling handy, building your own system is always another option. Regardless, indoors or out, there is a system to fit most anyone's composting needs.

CHAPTER 4

RED WORMS AND VERMICOMPOSTING

In Chapter 3, we referred to the use of worms in composting. Known as **vermicomposting** (*vermi* meaning "relating to worms"), these worms should not be mistaken for the huge night crawlers used for fishing bait—these are specialized breeds for composting. The worm variety that we'll be discussing here is *Eisenia fetida*, better known in the composting world as **red worms** or **red wigglers**. This species of worm is adapted to thrive in the decaying materials of compost, whether in bins or in nature. Also known as the tiger worm, red California earth worm and trout

While different cultures have access to different types of worms, the consensus is that vermicomposting is safe, effective, and here to stay! Photo by Find Your Feet under the Creative Commons Attribution License 2.0.

worm, red worms thrive in colonies. They are unquestionably the workhorse of the compost pile, especially in a vermicomposting system. While other breeds, such as the European night crawler and African night crawler, are suitable for composting, we'll be restricting our discussion to the red worm, as it is the most common and widely available variety.

Many composters and gardeners consider vermicomposting the be-all-end-all of composting—the best method for getting the highest quality compost. The reasoning behind this relies on the understanding that not all compost is created equal. While many believe that compost is simply taking something that was living and breaking it down until it becomes dirt, this is not quite accurate. There are differences between the two styles of composting. Basically, *composting* breaks down organic matter using the heat that is caused by microbial metabolism (along with the size of the pile or heap), while *vermicomposting* breaks organic matter through the use of worms and microorganisms. Unlike a compost pile, where high heat and low moisture are what makes the pile work, vermicomposting prefers cooler (but not cold) temperatures and higher moisture.

The soil that is produced through vermicomposting is a mixture of **worm castings**, which is any organic material that passed through the worm's digestive system, **mucus secretions** from their bodies, as well as the organic material itself. In other words, worm castings are "worm poop!"

The worms allow certain microorganisms to thrive while eliminating others. The resulting compost produces natural growth hormones and enzymes for plants, along with other things that the plant might need. Furthermore, with vermicomposting, you can continue to add new organic materials. Unless you have more than one bin, at some point you will need to begin to stockpile new organic material until you are able to start a new pile. Perhaps best of all, vermicomposting makes for the perfect indoor composting system, even for an apartment dweller.

That being said, vermicomposting is not without its drawbacks. First and foremost, the worms will need some care and attention. While they do not require the level of care that a pet or livestock needs, you will need to make sure their surroundings are moist enough for them; make sure that they are kept out of excessive heat (which can kill the worms) and excessive cold; and, depending on the type of set-up that you have, you *will* need to remove your worms by hand when it comes time to remove the compost from the bin (which might be a job that the kids would enjoy)! Finally, if you keep your vermiculture bin outdoors, the cold weather can limit your results (as low to freezing temperatures can slow and even halt your worms' activity).

|||||||||||||||||||||||||||||||||||||||||||||||||||||||||||||||||||||||||||||||||

Vermicomposting bins, like compost bins, should never smell. If they do, then there is an underlying problem, one which you will need to identify and correct.

|||||||||||||||||||||||||||||||||||||||||||||||||||||||||||||||||||||||||||||||||

Bins for vermicomposting come in all shapes and sizes, and can be purchased commercially (or else homemade). Red worms will live anywhere as long as their environmental requirements are met, and because the worms will usually live and eat within the top 6–8 inches of their bedding area, the bins do not need to be very deep.

The box/tray types of vermicomposting containment are:

Traditional: The most common and perhaps the oldest type of bin, these are most commonly constructed from plastic storage containers, although other plastics may be used, such as buckets (including plastic cat litter buckets), so long as they can hold at least two gallons.

Flow-Through System: Usually made from a wooden box, metal drum or plastic barrel (some even use a new, covered plastic kitchen garbage can), a flow-through system keeps the food and worms on top while the castings are harvested from the bottom. The bottom is left open, with doweling, pipes or wire placed above the opening. On top of this is placed cardboard, or possibly heavy paper—something that will prevent the worms and food from falling through. Some who use this system find that the worms will pretty much stay in the upper food area on their own. This is due to the fact that the bottoms are regularly cleared out of compost/castings, especially in a small system. Since there is nothing for the worms to burrow down into, they take it upon themselves to stay up top.

Stacked Bins: This method is great for those who have limited space, and is good for indoor use. Using the same bins or buckets (or trays) that we have discussed, bins are stacked one inside the other (the number of bins used is up to you). The stack then sits on a base, which is itself another bin. This bin will remain empty in order to collect moisture—your "worm tea," a prized byproduct. The top bin should have a tight fitting lid to keep moisture in.

Effective, clean and easy to store, this type of stacked set-up can be easily implemented in a variety of situations. Photo by Kafka 4 Prez under the Creative Commons Attribution License 2.0.

The upper bins have holes in the bottoms to allow the worms to migrate from one bin to the other, once the ingredients of their current bin or tray have been converted into compost/castings. After a bin has been emptied, it can be moved to the top of the stack.

Trays: Basically, trays function in the same way as shallow bins except that they are larger, aren't stacked and are normally utilized by commercial operations. Although this set-up *can* produce castings, this configuration's main purpose is to promote worm reproduction. Because it is a highly controlled environment, worms in a tray system are only given time enough to produce their cocoons. Once that happens, they are moved to a new tray to begin the process all over again. This system is more useful for those with some sheltered space.

Outdoor: There are a few outdoor vermicomposting systems; however, we will only be discussing two methods, as they are best suited to the backyard and limited space areas.

Buried bins are just as the name suggests. The bins are buried in the ground, level with the surface. This creates a sort of contained system. Although this system has a chance of worms making their escape, if conditions are properly kept (including keeping the worms well fed/maintained), this shouldn't be an issue.

Above ground bins are large bins that are kept above ground. If you are in an area of harsh or variable weather conditions, these bins should be insulated and/or temperature controlled.

While both buried and above ground bins are intended for outdoor use, remember that they require proper care and precautions. They may also not be viable year-round, depending on the climate in your area. That said, depending on exactly how cold it gets in your region, you may have some success keeping your worm bins out year round.

If you are in an area of mild cold and snowy winters (as opposed to extreme cold and snow), you may move your bin/tower up against the house, and build a wooden box outside of and around the bin. Then, using straw, foam insulation sheets, etc., insulate between the bin and the wooden box. If you have a couple of old windows, you can place these on top of the boxed-in bin at an angle. This will help to pull in a bit of heat, as well as keep the snow off of the top. Before you tuck your worms in for the winter, however, don't forget to provide them with lots of food and leaves to keep them happy and fed throughout the winter; remember, the more you open the bins, the more heat you will let out.

Along with the boxes, some have had success in putting their bin next to their dryer vent. While this will not keep your bins from freezing, if you live in areas that have cool, mild winters, this can provide your worms with a little extra warmth.

Finally, there are special heaters available to help keep worm bins warm. The best ones to use will have a thermostat which automatically controls the on/off setting and temperature. If you cannot find these heaters, you may also use seed tray warming mats (again with thermostat), and lay them on the worm bins. However, even with all of these options available to you, if you live in a region with severe winters, it is still best to bring the worms indoors; even with your best efforts, your worms may still freeze once it gets cold enough.

Locating Your Worms

By now you might be wondering: how do I acquire these worms? The truth is that red worms are pretty easy to find. If you have worm farms in the area, they'll almost certainly be able to help you; if not, they can be ordered online

Red worms, like the ones shown here, are the secret ingredient that makes vermicomposting so successful and effective. Photo by crabchick under the Creative Commons Attribution License 2.0.

or through ads placed by homesteaders in garden magazines. If you purchase a commercially made system, they may come with ordering coupons that you can redeem through the mail to receive your first worm shipment. All worms are hermaphrodites, having both male and female reproductive organs. This means that you do not need to worry about your "male to female" ratio in the bins; however, the worm cannot mate with itself, so it will still need a partner to reproduce. Finally, if you have friends with a few compost piles of their own, many times a properly kept pile will naturally attract red worms. If they have a large number of worms in their compost (or bins), they may be willing to share!

No matter what you decide, whether you purchase a pre-made system or turn it into a DIY weekend project, you shouldn't have any problems getting your little workforce together.

Sample Vermicomposting Plans

There are many styles of vermicomposting bins. The following pages show a basic plan for a weekend project.

You can also find links to more ideas on the Resources page.

Worm Composting Bin

Worms in the house? Yuk! But this composting system actually works! The worms stay in the box and eat household scraps, and the box gives off little odor. Worm composting can be done in apartment buildings or other homes with no yard space. You might try it in your school!

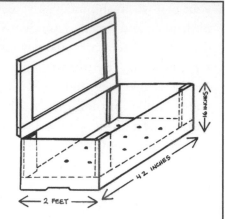

What You Need

Materials

- 1 4-x-8-foot sheet of ½-inch exterior plywood
- 1 12-foot length of 2 x 4 lumber
- 1 16-foot length of 2 x 4 lumber
- ½ pound of 16d galvanized nails
- 2 pounds of 6d galvanized nails
- 2 galvanized door hinges
- optional: 1 pint of clear varnish or polyurethane
- optional: plastic sheets for placing under and over the bin
- 1 pound of worms for every ½ pound of food wastes produced per day (Worms sold as fishing bait are best. Red worms are available from Flowerfield Enterprises, 10332 Shaver Road, Kalamazoo, MI 49002, 616-327-0108.)
- bedding for worms: moistened shredded newspaper or cardboard, peat moss, or brown leaves

Tools

- tape measure
- skill saw or hand saw
- hammer
- saw horses
- long straight-edge or chalk snap line
- screwdriver
- drill with ½-inch bit
- eye and ear protection
- work gloves
- optional: paint brush

Building a Worm Composting Bin

1. Measure and cut the plywood as shown, so you have one 24-x-42-inch top, one 23-x-42-inch base, two 16-x-24-inch ends, and two 16-x-42-inch sides.
2. Cut the 12-foot length of 2 x 4 lumber into five pieces: two 39-inch pieces, two 23-inch pieces, and one 20-inch piece.

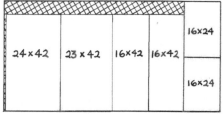

3. Lay the five pieces on edge on a flat surface to form a rectangle with the long pieces on the inside and the 20-inch length centered parallel to the ends. Nail the pieces together with two 16d nails at each joint.
4. Nail the 23-x-42-inch piece of plywood onto the frame with 6d nails every 3 inches.

5. Cut four 1-foot lengths from the 16-foot length of 2 x 4 lumber. (Save the remaining 12-foot piece).Take the two 16-x-42-inch pieces of plywood and place a 1- foot length flat against each short end and flush with the top and side edges. Nail the 2 x 4s in place using 6d nails.

6. Set the plywood sides up against the base frame so the bottom edges of the 2 x 4s rest on top of the base frame and the bottom edges of the plywood sides overlap the base frame. Nail the plywood sides to the base frame using 6d nails.

7. To complete the box, nail the 16- x-24-inch pieces of plywood onto the base and sides at each end.

8. To reinforce the box, make sure a nail is staggered at least every 3 inches wherever plywood and 2 x 4s meet.

9. Drill 12 one-half-inch holes through the plywood bottom of the box for drainage.

10. To build the frame for the lid, cut the remaining 12-foot piece of the 16-foot length of 2 x 4 lumber into two 45-inch pieces and two 20-inch pieces. Lay the pieces flat to form a rectangle, with the short pieces on the inside.

11. Lay the 24-x-42-inch piece of plywood on top of the lid frame so the plywood is 1 ½ inches inside all the edges of the frame. Nail the plywood onto the frame with 6d nails.

12. Attach the hinges to the inside of the back of the box at each end (on the 2 x 4) and the corresponding undersides of the back edge of the lid frame, so the lid stands upright when opened.

13. The unfinished box should last for at least five years; finishing the box with varnish or polyurethane, however, will protect the wood and prolong the life of the box. Two coats of varnish with a light sanding between coats should be sufficient.

14. Find a good location for the box. It can be placed anywhere as long as the temperature is more than 50°F (10°C). The most productive temperature is 55° to 77°F (13° to 25°C). Garages, basements, and kitchens are all possibilities as well as the outdoors in warm weather (not in direct sunlight). Make sure to place the box where it is convenient for you to use. It is wise to place a plastic sheet under the box.

Adding the Worms

Moisten the bedding material for the worms by placing it in a 5-gallon bucket and adding enough water to dampen all the material. Don't worry about getting the bedding material too wet because the excess moisture will drain off when it is placed into the composting bin. Be careful if you use peat moss because it will hold too much water. It is a good idea to put wet bedding material into the bin outdoors and wait until all the water has drained out (one to two hours).

Add about 8 inches of moistened bedding to the bottom of one side of the bin. In go the worms! Leave the lid off for a while and the worms will work down into the bedding away from the light.

Adding Your Wastes

Dig a small hole in the bedding and add your vegetable and fruit scraps. Then cover the hole with bedding. Small amounts of meat scraps can be added in the same way. Do not add any inorganic or potentially hazardous material such as chemicals, glass, metal, or plastic.

Maintaining Your Compost Pile

Keep your compost pile moist, but not wet. If flies are a problem, place more bedding material over the wastes or a sheet of plastic over the bedding, or try placing some flypaper inside the lid. Every three to six months, move the compost to one side of the bin and add new bedding to the empty half. At this time, add food wastes to the new bedding only. Within one month, the worms will crawl over to the new bedding and the finished compost on the "old" side can be harvested. Then add new bedding to the "old" side.

From *COMPOSTING: Wastes to Resources*, Cornell Cooperative Extension

Prepping the Bin for Your Worms

Now that you have your bin of choice, what's the next step? If you have purchased a commercial bin, it should include full directions on how to prepare it for your worms. For those who have chosen to build their own bins, here are some basic steps for setting things up. (Note that if you have built a bin with multiple trays, your preparations may differ slightly.)

While a multi-level vermicomposting set-up, or "worm tower," may be effective, it may also be more complex than is strictly necessary for a first-time vermicomposter. Photo by Mosman Council under the Creative Commons Attribution License 2.0.

The first thing you will need for your bin and your worms is **bedding**. Rather than using soil, you'll be using newspaper, cardboard or even peat. Newspaper is most commonly used; virtually everyone has some, and it's a great way to recycle (use black and white print only, as color may be toxic).

Tear or shred the newspaper into 1-inch wide strips. The shredded newspaper will provide the food, air and water that your worms need. If using cardboard, it should be torn into small pieces.

Next, you will need to moisten the strips. You can do this by placing the strips into a large container or garbage bag, adding water until the strips have the feeling of a damp sponge. Do not use chlorinated water; if your tap water has chlorine in it let a few jugs

Shredded newsprint and copy paper provides everything your worms need to do their thing!

or buckets of tap water sit out for a few days before use, as the chlorine gas will dissipate (usually after 24 hours). When looking to tell if the paper is too wet, take a handful of strips and squeeze. If only a few drops of water come out, the paper is sufficiently moistened. If more than just a few drops come out, the paper is too wet, and some dry strips will need to be added. If the paper is too dry, add a little more water. When you have the correct moisture in your strips, they are ready to add to the bin.

As you're adding the wet strips, break up any large clumps. When putting the strips in place, make sure to fluff them up, and avoid packing the strips in tightly. You want to give your worms room to move around! Add strips until your bin is about three quarters of the way full. Then, in order to introduce some beneficial microorganisms into the bin, add up to four cups of potting soil or outdoor soil; just make sure it is clean, with no fertilizers, chemicals, or additives.

Now, it's time to add your red worms. You may be wondering how many to begin with; if you have purchased a commercial kit, it often comes with a coupon included to redeem for worms, in which case you will be starting with that predetermined number. However, if you are ordering/purchasing worms yourself, start with about a pound; it should not take them long to begin to procreate, thus increasing your population. Your starting count also depends on your preference and the size of your bin. If you have hand-crafted your bin, you may need more worms.

So, now you have your worms. What's next? Some will let the worms sit for a day or two to allow them to work their way into the bedding and start feeding. Others will feed the worms at the same time as they are added. No matter which method you decide, the food should be chopped into small pieces and buried under the bedding. Fruits, vegetables and other food scraps may now be added; however, limit your addition of citrus, as the peels can create an acidic environment that the worms will not like.

Make sure to add absolutely *no* meat, bone, dairy or oil.

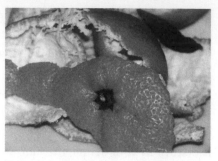

Citrus peels, like those of lemons and oranges, don't mix well with vermicomposting systems.

Your worms should be fed to half their weight. For example, if you average one pound of waste added to the bin per day, you should have two pounds of worms. To figure the weight of the worms, simply weigh them before you put them in the bin. Monitor the worms to make sure they are eating, and adjust your feedings accordingly.

To help keep your bin damp (especially if it is a small, indoor bin) place a few full sheets of dry newspaper on the bedding (the newspaper will function like a cover). Your worms will not survive in a dry environment.

This will also help if there happens to be any odor (of which there should be little or none), or if you are having issues with fruit flies. Replace as often as necessary if paper gets wet or fruit flies begin to increase in number.

Keep the bin at a temperature between 55–75°F (anything hotter or colder could kill your worms). Keep the bins away from any vibrations, as the worms will not respond well to frequent tremors. If your bin is indoors, keep it away from washers, dryers and other equipment that can cause vibrations that might affect your bin.

If the bin is indoors, keep it away from heat and windows. If your bin is outdoors, do not let it sit in the hot sun. Monitor the temperature of the bin and relocate accordingly. You may be surprised to learn that your worms breathe through their skin. Unlike most animals, worms do not have a "nose;" rather, it is their moist skin that allows them to breathe. For this reason, your worms cannot be allowed to dry out; if they do, they will die from suffocation.

Fluff the worm's bedding weekly to allow them enough air. If the strips feel like they are drying out, spritz them with water. If they seem too wet, add extra dry strips and fluff.

Vermicomposting can be a fun way to not only make compost but also acquire the much sought-after worm castings. Systems can be kept indoors or outdoors, come in any size you choose and can even produce a little "worm tea" (using the right

These bright red creepy-crawlies work hard to make sure that you can enjoy fine, high-quality compost. Photo by Red58Bill under the Creative Commons Attribution License 2.0.

systems). It is also a great form of composting for getting the kids involved, letting them create their own system to care for. Even if you don't have a garden, your houseplants will love compost resulting from these hard-working little worms!

CHAPTER 5

WHAT (AND WHAT NOT) TO COMPOST

. . . .

Up until now, we have been discussing the basics of composting, including some of the various containment systems and how they can be used on a backyard farm or homestead. Now, we will be getting a bit more into the nuts and bolts of the actual composting process, beginning with what and what *not* to compost.

What You Should Compost

According to the EPA, 20–30 percent of the municipal solid waste that ends up in landfills each year comes from yard and food waste, which should be composted. But does that mean we can throw anything in a compost pile? Or should we demonstrate more caution?

Of course, organic wastes are best for producing good compost. **Organic waste**, as it relates to composting, falls into one of two categories: **the browns and the greens**. Browns and greens here refer to waste products for compost which contain nitrogen (greens) and carbon (browns).

Greens are high in nitrogen or protein, and include:

- fruit and vegetable scraps
- breads
- pasta
- coffee grounds and used coffee filters

Greens are not only great for nitrogen—the worms love them, too! Also included in the greens category are alfalfa, clover and hay (although some dispute the practicality of using hay, due to weed seeds getting into the pile), other garden waste, grass and hedge clippings, seaweed, manure, weeds (past the seed stage, unless you want weeds taking over the pile) and crushed egg shells. Don't worry if something is moldy; it can still go into the compost pile. You can even compost old grains and spices the next time you clean out the cupboard.

Greens, or nitrogen-rich materials, are required for the composting process to proceed—but only in small doses. Photo by Steven Depolo under the Creative Commons Attribution License 2.0.

||||||||||||||||||||||||||||||||||||||||||||||||||||||||||||||||||||||||||||||||

Adding Lime to Compost

An old gardening trick is to add lime in the composting process; however, it is unknown what the benefits of this may be, if any. In fact, lime can reduce the nitrogen content of your pile. As a result, some gardeners will only add lime if there is a pH problem in the soil, adding it only to the finished compost right before applying it to the intended site.

||||||||||||||||||||||||||||||||||||||||||||||||||||||||||||||||||||||||||||||||

Browns, on the other hand, are high in carbon or carbohydrates, and include:

- paper
- junk mail
- paper egg cartons
- cardboard (in small amounts)
- dried leaves

Browns, the colloquial term for carbons, are essential as an energy source for any compost pile. Photo by Shopping Sherpa under the Creative Commons Attribution License 2.0.

Your pile can also include wood ash (when used sparingly; see the following box), bark and sawdust, shredded cardboard (in small amounts; put any excess in the recycle bin) and newspapers, corn and vegetable stalks, pine needles, straw, leaves and twigs (broken into small pieces to make the decomposition process faster).

Wood Ash

As stated, wood ash may be used in your compost pile, provided that it is used sparingly. Wood ash adds phosphorus, potassium, and micronutrients (such as manganese, iron, boron, zinc and copper) to your pile. It can also help to adjust a pile that is too acidic. However, wood ash is highly alkaline, which is why it needs to be used sparingly. Wood ash used in compost piles should come from a fireplace or even a campfire (as long as you are only burning logs), as opposed to the wood ash that comes from charcoal, or other wood sources (such as old 2x4s, crates, palettes and such), as these may contain chemicals which you do not want leaching out into your compost. That said you should also be cautious with ash from fruit wood, as many fruit trees are sprayed with chemicals throughout their lifespan.

If you raise chickens for meat and find yourself with an abundance of plucked feathers, you can add them to the pile as well (provided you chop up the plucked feathers first, as they can take time to break down). You can even add hair, as long as it is scattered throughout the pile and not allowed to clump.

If your pile is hot enough, you can add sod as well; however, if it is *not* hot enough, you will soon have grass growing in the pile. Chopped up leather can also be added into the pile or bin, as can chopped nut shells.

> Although it is neither brown nor green, but is rather neutral, you can throw rinse water and old beverages into your compost for moisture. Be careful not to add too much, though—you don't want your pile to get too wet (remember that your pile is also getting natural moisture from rain and dew).

What NOT to Compost

The things that you need to avoid putting into your compost piles or bins include:

- coal ash (as the sulfur and iron may end up damaging plants)
- color paper/cardboard (as they may contain toxins)
- diseased plants (if your compost pile is not hot enough, the disease will not be killed and you could end up with contaminated compost)

Stay away from inorganic materials; besides the risk of their containing toxins, many of them will not break down, either. You *can* compost wood, but make certain to leave out any treated lumber, as the chemicals wood is treated with can add toxins to your compost. Admittedly, some modern treatments leave out the toxins, so make sure you know which treatments your wood received before you add it to the pile. With this in mind, you should also steer away from all synthetic chemicals, as they can poison your compost (many synthetic chemicals will not break down at all).

Meat, bones, fish, dairy and fats can attract animals, cause odors in the compost (remember, your compost pile should always smell like fresh woodlot earth), and may overheat the pile. Rice in the compost pile can become a breeding ground for bad bacteria, so refrain from adding it to your pile.

Finally, don't dump the cat's litter box or dog piles in the compost. These can contain bacteria, parasites, pathogens and other things that could be harmful to humans (especially if the compost is used in food gardens).

||||||||||||||||||||||||||||||||||||||||||||||||||||||||||||||||||||||||||||||||||||||

Black Walnut Trees and Composting

Many say not to compost black walnut tree waste. This is due to the fact that it releases a substance called **juglone**, which can prevent growth in plants near the root zones of these trees. Juglone can kill blackberries, tomatoes, potatoes, blueberries and apples. However, some gardeners *will* compost walnut leaves, arguing that it is only the roots that are the problem. But even these brave gardeners will then test their compost before using it by planting a tomato seedling in it, to see what happens. In regards to black walnut tree waste, use your better judgment. It could work well, but then again, you could end up with contaminated compost that you can't really use for your plants.

||||||||||||||||||||||||||||||||||||||||||||||||||||||||||||||||||||||||||||||||||||||

Adding Compost Material to Your Pile

As far as the organic matter that goes into the pile, if you are gathering up outdoor waste, you'll likely want to gradually add it to your compost. Pile the material nearby and add as necessary. However, when disposing of kitchen scraps, you don't really want a pile of old food sitting on your countertop for weeks on end. Thankfully, there are ways to deal with this.

Using either a kitchen compost pail or a five-gallon bucket (with a cover) you can safely store scraps (which should be chopped up to save space) until you're ready to take them out. Depending on how much waste you produce, this could be once a day or once every other day. If you don't have room for storage containers of that size, you can also put scraps in a tightly sealed bag in the fridge for a day or two. Regardless of which storage method you use, they should be emptied as soon as possible to help alleviate any smells or fruit flies that may gather. As the five-gallon bucket lid may not be as tight fitting as the kitchen compost bucket, you can put some newspaper under the lid to help prevent fruit flies and odors. Another tip is to keep a little compost next to the bucket or kitchen bin; as you add new scraps, you can add a little of the compost into the bin as well. This will not only help with flies and odors, but will give the scraps a bit of a head start in breaking down before they even reach the main bin or pile. Whatever storage container you end of using for your scraps, it needs to be kept washed, or it will end up leaving a terrible smell in the house.

Don't think of your compost pile as a second garbage can. Keep careful track of what you add to your pile, and in what quantities. Photo by Tim Jewett under the Creative Commons Attribution License 2.0.

When first starting to add to your compost pile, you should alternate between layers of brown and green material. Every few weeks, turn the pile; once the composting has actually begun, you can just dump in the new additions instead of continuing with layers.

But how much of each type should you add? Some say equal amounts of each, while others prefer twice as much brown as green content. Having a bulk of browns will allow for the penetration of oxygen, all-important for nourishing organisms. However, if there is too much nitrogen, decomposition will be slow and you might end up with a smelly, dense pile. Basically, it comes down to what is working well for you. And remember: when adding to the piles, spread out what you're adding; make sure to chop or shred it, and avoid clumping for a faster break down.

|||||||||||||||||||||||||||||||||||||||||||||||||||||||||||||||||||||||||||||||

Composting Leaves

We briefly discussed putting leaves in your compost pile, but when fall rolls around and you have far more leaves than what you can conceivably use in your compost pile, what should you do? The answer: instead of adding them to the pile, compost the leaves on their own.

Choose a shaded, well-drained spot for the leaf pile. Place a layer of leaves and a layer of dirt until the leaves are all contained within the pile (don't pack the pile too tightly). A good pile should be four feet in diameter and three feet high. The leaves will most likely take up to six months to break down into compost. Keep in mind that leaf compost in and of itself is not nutritious enough to work on its own as a fertilizer, so it is used more as an organic soil conditioner.

|||||||||||||||||||||||||||||||||||||||||||||||||||||||||||||||||||||||||||||||

Composting Livestock

It may not sound pretty, but death is a fact of life on a farm. Whether as a result of culling or natural causes, farmers will eventually find themselves with an animal carcass, large or small, that needs to be disposed of. Disposal options include rendering or incineration, both of which require the animal carcass to be picked up, and can be expensive. (Incineration/cremation can work well for small animals, but as most farms do not have on farm facilities, this process may be very expensive, especially in sending large animals out for cremation.)

Rendering

Rendering is not as readily available as it was. Even in areas where rendering *is* still an option, its convenience (especially for larger animals) is offset by the expense of having the body hauled away. Plus, you will likely need to be able to "store" the dead animal for as long as a few days or more before someone will be able to come out and pick it up. If it happens to be an especially hot summer week, you may be dealing with odors, flies and possible bio-security problems (along with irate neighbors who do not understand the process). Due to current rules and regulations, some rendering plants may not be allowed to take animals that have already died. And even if you do have a downed animal that is still alive, but which cannot walk, you may have further problems dealing with how the animal is handled by the rendering plant, which may not be the most humane (depending on the company).

Burial

You might think that the obvious, more common solution would be burial. You may then be surprised to learn that some areas frown on burying animals, even when it's done on one's own property. Many areas even have strict laws and ordinances in place preventing you from doing so. In these cases, being able to compost your animal carcass can be quite handy.

If you *can* bury on your property, there are things you need to consider. Stay away from wells and ground water to prevent contamination. Be sure that you are burying the animal deep enough; otherwise, you may have flies, odors and other animals coming in to try and dig the carcass out. Improper burial may also risk spreading disease. And even if burial isn't a problem, there may be reasons that you *can't* bury your animal: you may not be able to dig a deep enough hole; the ground may be too wet or frozen; the area where you can bury might be too near a well, risking future water contamination, or any number of prohibitive reasons.

This is where composting can be useful: composting livestock can save money, as you eliminate the need for rendering or incineration services. Composting gives you the best of both worlds—you eliminate the disposal problems of dead livestock; animals that are downed can be handled humanely right on the farm; you eliminate having to work around a third party; and you save a lot of money because of it, in addition to the satisfaction of having gone full circle with your animal. There's also nothing wasted; most animals that will be composted are those who have died due to circumstances where they most likely would not be suitable for consumption.

Composting livestock also minimizes the risk of contaminating the water table through burial, as well as other possible contamination problems. Some areas even offer funding to those who compost on-farm, with compost areas being eligible for grants or incentives as a farm improvement. Through composting, your animals get to have one final job on the farm, and you are able to create some good compost for your use.

The Process of Composting Livestock

Composting livestock, in theory, isn't much different than composting other organic matter. And, unless the animal died from a disease caused by spongiform pathogens, all livestock can be composted. You need the same basic knowledge of temperature, time and pH balance as you do with regular composting. And, as with composting other organic matter, you don't need any special building for composting livestock—although you could build a huge covered "bin," actually a three-sided shelter with a roof which, like the composting bin, helps protect the pile from too much water.

The size of the pile will depend on the size of the animal. A horse will need to be in a much larger pile than a few chickens or turkeys. Note that small animals, such as poultry, may be composted in groups, instead of in multiple small piles.

For the composting process, you will first need to select a site (keeping in mind that you will need to be sure that it is in an area that is easy to manage, as it will have to be properly maintained). A proper site includes one that is kept away from wells, streams, rivers and lakes. You will need to be able to manage any runoff so that surface and/or ground water is not contaminated. The site should also be on high ground, ideally in a flood-free zone.

If you have neighbors, you will want to keep the site out of view, as some may not appreciate witnessing the transport of animal carcasses or the composting process. Make sure that the site is out of your way as well, as the last thing you want is having to work around the pile because you decided to place it by the barnyard or too near to an equipment area (especially if you are composting large animals such as cattle or horses).

Finally, before you begin contact your local extension office to confirm whether there are any state or local regulations you will need to follow.

Once you have secured a proper site for the composting, you will need dry wood mulch (such as chips, sawdust or shavings), which is necessary to keep odors minimized. Plus, the dried materials will absorb the fluids from decomposition. If you don't have any (or enough) wood mulch and don't want to purchase it, then you can use other organic waste that you may already have on hand. Just keep in mind the carbon to nitrogen ratio (25C:1N). You can also compost right on the grass; however, composting on more hardened, packed ground will avoid the muddy mess you might see during the rainy or thaw season.

To begin the process, you will first need a bed of mulch on which to lay the animal. This should be a 2–3 feet deep bed on a flat surface; the larger the animal, the deeper the bed will need to be. The animal should then be placed on the bed and covered with 4 feet of wood mulch (or whatever dry organic matter you are using). Make sure that the animal is covered; if you can heap the material into a cone shape, it will help to shed water and encourage the bacteria necessary for the composting process to be successful. The animal should also be surrounded on all sides by at least 2 feet of mulch. You also may want to puncture the rumen (the first and largest division of the stomach of a ruminant, such as a cow, goat or sheep) to prevent bloating and help speed up decomposition.

An animal properly buried in a compost pile will produce no smell, and won't attract other animals looking to scrounge. In about three months (for a large cow or horse; around 30 days for something like a broiler hen), composting should be complete, with only a few bones left. You can tell when the compost pile is nearing completion; once the beneficial bacteria have completed their work on the animal's carcass, breaking down the chain molecules into water vapor, the "cone" will begin to fall in. If you find the pile settling, but composting is not complete and the animal carcass is becoming visible, add more dry materials as necessary to keep odors and pests at bay.

When composting livestock, keep in mind that the carcasses are dense—they're high in nitrogen and contain 60 percent water. As a result, a drier, more absorbent pile develops than one sees with a regular compost pile. High carbon materials are necessary to surround the carcass in order to balance essential nutrients and provide the correct environment for microbial growth. Due to the high water content of a carcass, these materials should have at least a 50 percent water weight. Too much moisture and you will have ammonia, as well as methane and/or nutrient-rich leachate, both of which are significant sources of odor (these are caused by limited oxygen). On the other hand, having too little moisture will limit the microbial metabolism, in turn slowing decomposition.

A simple rule of thumb for determining whether or not your material has 50 percent water weight is to pick up a handful of raw material and squeeze it tight. If it sticks together slightly, leaving only a few drops of water behind, you're in good shape.

The completed compost should have no tissue remaining in it, and should be odorless. Any bones left should be brittle, and the compost itself should be a rich brown or dark brown with a good, earthy scent. Texture-wise, it should be crumbly, but with the ability to hold moisture. Although this compost may be low in nutrients, you can mix it in and use with other nutrient-rich compost (such as you would do with leaf compost) or use it as a filter by laying some of the compost over a new pile, to help with odor reduction.

Although it is unclear as to whether composting kills all disease-causing organisms, viruses are usually more heat sensitive than bacteria or fungi, and typically become inactive between 122–144°C. As most compost piles reach these temperatures, there is seemingly little significant threat of a virus spreading. A properly managed livestock compost piles should not increase any risk of disease on the farm.

As a side note, I have used livestock composting myself (with a rather large pet dairy goat) and found it to be odor free, and no more difficult to do than regular composting. If you have the proper space to do so, it really is the way to go.

So, whether you're composting food scraps in the backyard or livestock in the field, following the basic steps of what you should and shouldn't add to your pile will keep you on the path to success in composting!

CHAPTER 6

CARBON AND NITROGEN RATIOS

I n the last chapter, we mentioned the importance of the carbon to nitrogen ratio in composting. Now, let's get a little bit more in-depth in regards to keeping your pile properly balanced.

As we discussed earlier, the carbon/nitrogen presence is very important to the composting process, as it affects the decomposition of organic matter—**carbon** serves as a source of energy, while **nitrogen** builds cell structure. However, these two elements need to be in the correct ratio in order to work effectively. The carbon/nitrogen ratio, or **C:N ratio**, is the relative proportion of the elements in the pile. For example, if something has 20 times as much carbon as nitrogen, the C:N ratio is 20:1.

We have already discussed browns and greens in Chapter 5, and their great importance in composting. For our present discussion, we need to understand that the **browns are carbons** and the **greens are nitrogen**. The two other requirements for composting (which are just as important as the C:N ratio) are water, as the moisture helps to break down organic matter, and air.

The best ratios are in the area of 25–30 parts carbon to 1 part nitrogen; in other words, a 25:1 or a 30:1 ratio. These ratios result in hot, quickly composting piles. If the C:N ratio is too high (if you have too much carbon as compared to nitrogen) decomposition will slow, as the pile will be cooler. Too much nitrogen, and ammonia gas and odor will form.

The decomposition of organic matter—its rate and its effectiveness—is affected by the relative presence of carbon and nitrogen. This C:N ratio indicates the relative proportion of the two elements; for example, a material with 25 times as much carbon as nitrogen is said to have a C:N ratio of 25:1, or put more simply, a C:N ratio of 25. This ratio should be considered as an estimate, or a working ratio; there may be some traces of material present (carbon, in particular) which is configured in such a way that it does not participate in the composting process whatsoever. These trace amounts are usually not included in calculating the C:N ratio (although you may notice them, in the form of the debris you clean from your finished compost!)

The reason that this ratio is important to decomposition is that the organisms involved in decomposing organic matter use carbon as their source of energy, and nitrogen as raw material for building cell structure. In finding an ideal ratio, it's important to bear in mind that these organisms need more carbon than nitrogen. However, if there is too *much* carbon, the process slows. The organisms expend the available nitrogen before finishing off the available carbon, and the balance falls apart. When this happens, some of the organisms die; others form new cell material using their stored nitrogen. In doing so, more carbon is burned, reducing the overall carbon amount while using only recycled nitrogen. However, decomposition under these circumstances naturally takes longer (these situations tend to occur when the initial C:N ratio is above 30, or 30 parts carbon to 1 part nitrogen). Organisms use about 30 parts carbon for each part of nitrogen, so an *initial* C:N ratio of 30 would promote rapid composting, and would provide some nitrogen in an immediately available form in the finished compost. However, the optimum value is actually a range, going from 20–31.

Another problem that results from an improper C:N ratio—again caused by having too much carbon and too little nitrogen—is that the microbial cells involved in the decomposition process will draw nitrogen from wherever it can find it—including the surrounding soil. Referred to as "robbing" the soil of its nitrogen, this has the adverse effect of depleting the nitrogen available to the bacteria naturally present in the soil, meaning that they can no longer play their role in fertilizing the soil. It may also come into play later, when there is no longer sufficient nitrogen for use in the life-cycles of your soil bacteria.

A C:N ratio of 20 is about as high as you want to go, to make certain that there is no danger of your soil being robbed of its nitrogen. Remember—the actual ratio may be still be higher—there may be carbon sources that do not participate in the decomposition process; however, so long as all active carbon sources together don't push the ratio over 20, you should be fine.

The C:N ratio is a critical factor in composting to prevent both nitrogen robbing from the soil—and conserving maximum nitrogen in the compost. In cases where the amount of available nitrogen outweighs the available carbon, the excess nitrogen is processed by the organisms and excreted as ammonia. This release of ammonia into the surrounding atmosphere produces a loss of nitrogen from the compost pile, and should therefore be kept to a minimum.

Composting time increases with C:N ratios above 30–40. If the amount of "unavailable carbon" is negligible, the C:N ratio can be reduced by bacteria to as low a value as 10. Fourteen to 20 are common values, depending upon the materials used to create the pile.[1] Likewise, a pile with a ratio in the area of 4:5 would result in slow composting. Organic ingredients in a pile with this ratio

---

1   http://whatcom.wsu.edu/ag/compost/fundamentals/needs_carbon_nitrogen. htm, Washington State University, Compost Fundamentals, Whatcom County Composting

(and letting nature take its course) would take up to a year to fully break down. Slow composting such as this usually fails to reach sufficiently high temperatures to kill any diseases that the organic matter may have had before it was put into the pile. It will also likely fail to kill any weed seeds, and may even have a bit of an odor. A hot pile using chopped organic matter (possibly with a little microbial inoculate, for example by adding fungi, and frequent turning) can be ready in 1–3 months. It will be odorless and hot enough to kill those unwanted seeds and disease. (There *are* some diseased plants or animals which should not be composted. If you are unsure, check before you begin the composting process using these items.) However, you *do* need to be vigilant with a hot pile. If the nitrogen becomes too high, your pile may get too hot, and in turn will kill off the necessary microorganisms in the pile, as well as possibly produce an odor.

A proper pile with the correct C:N ratio should begin to "cook" within a week. When the pile begins to cool, turn it to reactivate the carbon and heat up the pile again. Keep tabs on your pile; the best way to make sure the pile is properly composting is by keeping the pile moist. Test the pile's moisture by

It's important to keep your eye on your pile's temperature, as it's one of the best indicators as to your pile's C:N balance. Photo by Scot Nelson under the Creative Commons Attribution License 2.0.

digging at least a foot down into the pile. Add water if necessary (especially during periods of little to no rain), but do not let the pile become soggy.

Keep in mind that while 25:1 or 30:1 are considered ideal C:N ratios, actual ratios will vary due to a number of factors, including the organic materials used.

Some C/N ratio examples include:

- 19:1 (Grass clippings)
- 20:1 (Manure, old)
- 25:1 (Veggies, scraps)
- 80:1 (Straw)
- 175:1 (Newspaper)

There are numerous charts online for C:N ratios, some of which are listed in the Resources section of this book.

Trying to figure out (and remember) the correct C:N ratio and maintaining it in your pile may seem daunting; however, with all of the resources available online, there are plenty of quick references for use in keeping your pile's balance correct. Your area extension office or nursery should also be able to provide assistance in maintaining the proper ratio for your compost, often times even more so than the "big box" stores. While the employees at larger-scale nurseries are usually very knowledgeable about the products they carry, in terms of providing helpful advice for first-time composters (drawn from personal experience) places like the smaller nurseries and extension offices can't be beat!

CHAPTER 7

COMPOSTING PROCESSES AND STAGES

Much of our role in composting involves creating the ideal conditions for composting to occur. At its core, composting is a natural biological process, in which organic waste is broken up into humus. It is this all-natural recycling process that in the end produces the dark, nutrient rich soil we know as compost.

As we've discussed, successful composting requires organic waste, soil (if you so choose; adding some already composted soil to the mix may speed up the composting process), water and air. Temperature and moisture are both indicators that will assist you in monitoring and helping to control the process.

The proper conditions necessary for composting are:
- Plenty of air (through turning the pile)
- Enough water (the pile should be moist, but not sopping wet)
- Proper C:N ratio
- Small particulate sizes (large pieces of organic matter should be broken up into smaller pieces)
- Enough soil to provide the microorganisms necessary to facilitate the process (optional)

There are two types of processes that may be used in composting (whether in piles or bins). Most home composters use the **aerobic method**. This method uses air, moisture and heat to decompose organic matter; this is the method we've been primarily discussing thus far.

The second process is **anaerobic**, in which composting is done without use of air or heat. The pile is basically left to sit undisturbed until it completes its process. This makes for a longer composting period; both air and heat are used to speed along the decomposition process.

A wire bin like this one allows for more oxygen to reach more of the pile, allowing for greater aeration and aerobic activity. Photo by Nancy Beetoo under the Creative Commons Attribution License 2.0.

The aerobic method, the most common method used, also involves billions of microbes. Most are bacteria that love the hot, moist conditions of the aerobic method of composting. They eat, grow, reproduce and die; the energy that they expend by doing so is what is responsible for the temperature changes in the pile. As the pile cools down, you can reactivate it by turning it; this redistributes the microorganisms, and aerating begins once again, allowing for a sped-up decomposition and renewed microbial succession.

Compost Accelerators

If your pile isn't working fast enough for you, or you just want to speed up the process, you can either purchase a commercial accelerator, or else make one yourself!

Using a five gallon bucket, mix together the following ingredients and pour the result into your pile:

- 6 ounces of beer (the yeast acts as an accelerator)
- ½ cup ammonia (adding extra nitrogen helps speed up the breakdown process)
- 2 gallons warm water
- 12 ounces non-diet soda (provides sugar to feed the microbes)

If you don't want to use the liquid mix or a commercial product, save a bit of compost from any past piles and add a thin layer of finished compost to your organic materials. This introduces the microorganisms needed to speed up the process. Don't have any compost left? Not a problem, as top soil should accomplish the same thing. Water it to keep it moist, but not soggy (this cannot be stressed enough), and turn regularly (every 7–10 days) to keep it aerated.

Stages of Composting

So, we know that composting is a process that we can slow down or speed up, but what are the actual stages of the composting process?

The composting process consists of four distinct stages:
1. Mesophilic
2. Thermophilic
3. Cooling
4. Curing

There are a wide variety of bacteria that live in compost piles. In the beginning, most of the activity is carried out by the **psychrophiles**, which survive only in very low temperatures. They will begin to oxidize or burn the carbon in the pile. This in turn releases heat and nutrients (amino acids). This is only in the early stages, and is typically a very subtle early stage, and thus not often noticed.

When the pile gets too hot for the psychrophiles, the mesophiles take over. **Mesophiles** thrive in the medium temperatures seen early on in the composting process. Bacteria in the **mesophilic stage** usually colonize the heap within a few days, at which time the heat begins to increase. Mesophiles love food scraps and are important in the composting process. The bacteria also work when new organic matter is added to the pile, and do the most work in the composting process, eating everything in sight and generating heat over 100°F.

When the temperatures get too high for the mesophiles, the mesophiles will exit, and the process is continued by bacteria known as **thermophiles**, as they can survive in the very high temperatures of the active compost pile, up to 200°F. Living only 3–5 days, thermophiles spend their lives raising the temperature of the pile high enough to kill any seeds or germs that may be residing in the pile, all the while generating **humic acid**—a naturally occurring compound which is created through the biodegration of dead, organic matter and contain a number of trace minerals. These are the same bacteria that are responsible in cases of spontaneous combustion of hay, as well as very large, very dry compost piles. (However, the typical home compost pile is normally not large enough for this to be a concern.) If the heap gets too hot, it will need to be turned to cool, giving the pile one more round of the thermophiles working towards a faster decomposition. However, this stage happens quickly, and can last days or even months.

Both of these types of microorganisms (mesophiles and thermophiles) are common throughout nature, and are frequently found on food waste, garbage, manure and human waste (humanure), not just in compost piles.

As the temperature begins to stabilize, the ambient temperature will begin to set in. During the cool down stage, the pile gradually begins to drop in temperature. The mesophiles and psychrophiles become active once again, in what is referred to as the curing stage. At this point, worms and insects (among others organisms) will take up residence in the pile. Having completed all four stages, the process is complete and the compost is ready for cleaning and use.

Composting manure allows you to take advantage of the naturally occurring mesophiles and thermophiles in nature to produce rich compost.

CHAPTER 8

CLEANING AND USING COMPOST

Now, you've made your bins, you've collected your organic waste, tended it as necessary and waited patiently for the process to complete. And now you have it: your dark, rich compost. Although you could just scoop it out of the bins and use it now, it is best to clean it first. Cleaning compost refers to the removal of anything that didn't break down in the process through sifting. Seeds, small pieces of sticks, and any other lumps you might find should be broken down and discarded before using your compost.

As far as sifting the compost from a pile or a bin, the jury is still out as to what method is preferable. Some gardeners will sift it all, and are left with nice, fluffy compost. Others won't sift at all, and will just scoop compost directly from the pile when they want to use it. Still others will split the difference, depending on the task at hand. If they just need some compost for a hole when planting, for example, they won't sift. However, if the compost will be visible—for instance, if it will be lying on top of the garden or in a container plant—they will sift it to make it look a bit more finished and neat. The choice is yours, and is primarily a matter of personal preference.

To sift compost, you can either purchase commercially-made sifters, or else make your own using either a wire mesh hardware cloth (½ inch), chicken wire (for a coarser sift) or anything else that works well for you. Mount the wire on some type of frame; I've had considerable success using old window screens, still in their frames.

What you use to clean your compost is up to you, but chicken wire and old window screens make for excellent, large-scale sieves. Photo by Sustainable Sanitation Alliance under the Creative Commons Attribution License 2.0.

Cleaning Vermicompost

When it comes to vermicomposting, you will definitely need to clean the resulting compost (by removing the worms). However, you probably will not want to sift it; you don't want to hurt your valuable worms. If you are using stacking bins or stacking trays, most of the work will already be done for you, as when one level is done, most of the worms will automatically migrate up to the next level, where there is new food and bedding available for them. Once they've left the now composted materials and castings in the lower tray, just remove the tray, dump out the compost into another container, and remove any worms that may be left behind. Once emptied, replace the tray on top of the stack.

However, if you aren't using a stacking system, you will need to look for ways to remove the worms from the compost. This can be accomplished in a few ways. You could sit with your bin and pick out the worms, one by one, but this is a tedious, time consuming job. You might be able to get the kids to do it for you, but there *are* easier ways.

Some have found success with pushing the worms and their compost to one side of the bin, then put new bedding and food down on the now empty side. The worms should then migrate to the side with the new food bedding, beginning the process all over again. Once the worms are on the other side, scoop out the compost and castings.

Worm casts, shown here, are a big part of what makes vermicompost so potent and effective. Photo by Red58 Bill under the Creative Commons Attribution License 2.0.

Another method is to use light to remove the worms. For this method, you will need:

- A large sheet of heavy plastic
- A lamp (optional if in daylight or a bright room)
- Two buckets (one for the worms and one for the vermicompost)
- New bedding

Place the plastic sheet on the floor or on a large table and dump the bin's contents onto the plastic. Make at least nine cone-shaped piles from the compost. As you do this, you should be seeing a lot of worms at first, but they'll disappear as the light forces them to move deeper into the compost. If the room is dark, use artificial light; however, if it is daylight or you're in a bright room, the artificial light *may* not be necessary. Once the worms are all "gone," go to the first pile that you made, and carefully remove the surface of the pile. If the worms have not yet moved, wait for about 10 minutes or so, then come back and check. Do the same with each pile that you made. Don't forget to put that surface compost into your bucket. By the time you are done with all the piles, the worms should have once again migrated deeper inside their respective piles, and should be have moved again. If not, wait for a few minutes. Repeat the surface removal until the worms can be found in a heaping mass at the bottom of the pile. Take the worms and place them in new bedding to begin the process all over again. If you find you have a lot of worms, this is a good time to begin another bin; or, if you like, you can give away or sell the excess.

Using Your Compost

Now you have a pile of wonderful black dirt in your backyard that you created over the past few months to a year. Now, you need to figure out what to do with it! Not to worry: compost is one of the easiest things in the world to use, despite its well-earned place as one of the biggest boosts for your plants and garden. Consider all that compost can do for you: improving the soil as it adds nutrients; feeding the earthworms and microbial creatures; protecting plants from disease; and supplementing the soil's organic matter. Not only can it help to increase the amount of water your soil can hold (as compost charac-

Affectionately known as black gold, compost can work wonders on any garden—it really is worth all the effort you put into it! Photo by wisemandarine under the Creative Commons Attribution License 2.0.

teristically holds more moisture than plain soil), if your compost is especially high in nutrients, you might even be able to eliminate the need for fertilizer. All of this from something that began as a bit of your household garbage!

First, know that you should use only completed compost; not

50 percent ready, or even 75 percent ready. It should be 100 per-cent completed, with no food scraps remaining. You might have a few stems or parts of things remaining that just didn't compost all the way down like you thought it would, but in this case, you would clean the compost (as we discussed before) using some type of screen or sieve. Once you have nothing but smooth compost, it is ready to use. But how?

Perhaps the easiest way to use compost is as a mulch, spreading a thin layer of the compost over the top soil. Don't worry about blending the two together—Mother Nature will take care of that through opportunities such as weather, settling and worms.

Your compost will also make an excellent nutrient dressing for your lawn. Simply spread a layer of compost over the grass, especially in areas where you think extra nutrients are necessary. Don't bother to rake it in or dig it in, as once again Mother Nature will take care of and the task of settling it in for you. It would be advisable, however, not to walk on these areas until the compost has settled in, so that you don't end up with a trampled mess, as this would make it difficult for the soil to settle down into and around the grass.

If you need an area to retain moisture better than it does (such as in cases of sandy soil), try using some of your compost. As an organic material, compost holds moisture well (about 20 times its weight), and as a result can help soils to retain extra moisture as well. For example, for every 1 percent of compost (organic matter) added, an additional 1.5 quarts of water per cubic foot of soil can be held.

If you have some trees that need extra nutrients, compost will work great for giving them an extra kick! If you haven't planted the tree yet, after you dig the hole, mix a bit of compost in with the soil that you have put aside to refill the hole after you have set the tree. If the tree is already established, determine the root area, and simply spread a layer of compost over the root area. When doing this, you do not need to be concerned about mixing the compost into the soil, as nature will do that for you. However, you do need to be careful about avoiding the base and the trunk of the tree. With any mulch, whether it is compost or coconut or wood, if you pile mulch up against the trunk of a tree (or bush), moisture can build, which could cause the trunk to rot (or else introduce insect problems to the tree).

When tilling your garden, consider spreading the compost over the garden area before you till. This will allow you to incorporate the compost into the soil as you go along with the tiller, basically taking care of two jobs at the same time. Of course, if for some reason there are only certain plants that you wish to use your compost with, you may add it to only those specific places as you plant your seeds or seedlings.

Incorporating compost into your garden successfully may take a bit of forethought—but the results make it absolutely worth it. Photo by hardworkinghippie under the Creative Commons Attribution License 2.0.

If you are a container gardener, don't worry:—you can still compost! Completed compost makes an excellent potting soil due to the way it holds moisture, while still draining well. The compost will normally be incorporated with perlite, vermiculite or coarse sand,

Completed compost (in a sturdy compost bin) together makes up the perfect set-up for a container gardener.`

but can also be incorporated with regular potting soil. In small pots, however, I have used compost only, without incorporating anything else. And don't forget the compost tea; with its richness in micronutrients, it is also great for container use.

Herb gardens? The addition of compost will give you leafier and healthier herbs, as well as excellent vegetables, especially when used with root vegetables and potatoes. Make sure to mix the compost in a few months before you plant the carrots or potatoes, in order to assure good growth; too much nitrogen will promote greenery growth while inhibiting root/tuber formation.

Yes, using your compost really *is* that simple! No matter how you decide to handle your finished compost, you will find that you now have the "black gold" that all gardeners hold dear, *and* you'll be ready to begin the next batch!

CHAPTER 9

TROUBLESHOOTING

My compost pile is too wet (due to rain).

If rain is drowning your pile, try turning it and covering it loosely with a tarp.

My compost is dry.

Your compost should feel wet, like a damp sponge. If it doesn't, and is instead dry, water your pile (a sprinkler works great) before aerating the pile. It should begin to heat up again in a relatively short time. If necessary, add some extra nitrogen-rich material (what we refer to as greens in Chapter 5) for heat.

My compost is damp and smells right, but is not heating up.

If your compost pile seems to have the right moisture and smell, but no heat, then the C:N balance is probably off. Try adding more nitrogen rich material (greens).

There are bugs in my compost pile.

This usually won't be a problem, as most will not hurt the pile. When scooping out what you need, spread the compost out on paper and lay it in the sun before use, and let the insects scurry away.

My compost pile has ants.

If there are lots of ants in the pile, it is probably too dry. Moisten the pile and then turn it, which should disrupt their nest. Allow them some time to leave; if they don't leave, don't worry—it really won't be a problem.

My pile is attracting flies.

This is most likely due to any food waste that you are adding to the pile. Cover the food waste with nitrogen-rich material (greens), such as soil or compost. However some insects are beneficial, so before you decide to get rid of them, make sure of what you have.

Certain pests, like the garden soldier fly, may *seem* like pests, but are actually beneficial. Soldier fly larvae are extremely adept at breaking down green materials in compost piles. Photo by John Tann under the Creative Commons Attribution License 2.0.

My compost has plants.

If you find plants in your compost pile, the pile probably didn't get hot enough to kill the seeds. But the end result is quite simple to deal with: if the growing plants are weeds, simply pull them. If you find that they are usable plants (like vegetables, fruits or herbs), simply transplant them to a container or garden space.

Compost will occasionally see these types of unwanted guests. Simply yank them out, and keep a closer eye on your pile's temperature moving forward. Photo by glacial under the Creative Commons Attribution License 2.0.

My compost smells like rotten eggs.

If your pile has this odor, then there is not enough oxygen and it is too wet. You need to aerate (or turn) the pile and add more nitrogen-rich material (greens).

My pile has an ammonia smell.

If your pile has an ammonia smell, you have too much nitrogen. Add more carbon-rich material (what we refer to as browns in Chapter 5).

Animals are getting into my pile.

If you are having wildlife problems, such as raccoons, bears or other "visitors," or if the neighborhood pets are getting a bit too curious, you may need to bury your kitchen scraps deeper, as this is likely what is attracting the animals. You could also try mixing the scraps with soil or compost before burying. Are you trying to compost meats? Remember that this can attract animals to your pile, which is one of the reasons why composting meats isn't recommended in a garden compost pile.

However, if the problem persists even after these steps, you may need to consider investing in an animal-proof covered bin.

Can I purchase red worms to add to my outdoor compost pile?

Yes, you may; however, if you have a good, healthy heap that is housed directly on the ground, with a little patience worms will usually appear naturally and make your compost their home. If you need to purchase red worms, you may be able to purchase them locally. If not, there are many online sources that offer red worms. You might want to check shipping restrictions, however— the weather and/or seasons may have an impact on survival during shipping, and some places, depending on where they are located, may not ship at all during certain times of the year.

My worms died!

This happens. For some, it can take a few tries to get the correct balance in the worm bin. This is usually due to improper bedding, or bedding material that is too dry. Purchase some new worms, avoid any previous mistakes, and try again!

My worms are all in a ball. Why?

Most likely, the bin is too cold for the worms. If the weather is changing, either move the bin to a warmer area, or else bring it in the house for the winter.

Can I build my compost pile in the winter?

Yes! You may add to or build a new compost pile in the winter. Keep in mind, though, that due to the cold the composting process will slow down and even stop, depending on the temperature. However, your pile will be ready to continue on as soon as the weather warms up.

Can I compost over the winter?

If you are in an area that gets cold winters, it is possible to keep your pile active and "working," at least until the coldest part of your winter sets in. However, you *will* need to insulate the pile, which can be achieved through a generous covering of straw, leaves or newspaper (I prefer straw; should you want to get at some of the compost, straw is easier to push aside with your hands) and covering with a dark tarp. This may help the bacteria stay active until it gets extremely cold, at which point the composting process will stop.

Also, if you are using a movable bin, relocating the pile to a warmer area of the yard (or even nearer to the house) may give your pile more warmth in the winter, as well as grant you easier access—especially if you live in an area with heavy snow. Believe me, you will quickly become tired of traipsing out to the bins every night!

Continue to maintain your C:N ratio. It might help to chop up the material you want to compost to help aid in the composting process, especially through this slow time.

What if I am in a dry/drought area? How will that affect my compost?

If you are starting your compost pile in a region that is naturally dry or experiencing a drought, there are a few things you should do. It will help if you can create your pile in something that will help you retain water, such as a plastic barrel or bucket, to help with evaporation problems. Each layer you add should be soaked until it is quite damp (like a wrung out cloth). When rain is expected, create a bowl or crater-like formation in the center of the pile to help collect whatever moisture you can from Mother Nature. After the rain, you may want to check to make sure that the water hasn't puddled; if it has, you might want to mix it into the pile a bit more evenly.

Can I store my completed compost?

Yes, you can store compost. Ideally, compost is meant to be used fairly quickly, in order to take full advantage of the nutrients. However, if for whatever reason you find yourself in the position of having to store your compost, there are a few ways to do so:

If your pile (or heap) is right on the ground, you can simply cover the finished compost with a tarp or a plastic sheet. This covering will protect from the excess moisture of any precipitation, while still allowing some humidity to keep your pile moist. A pile kept on the ground will also still allow those beneficial worms to find their way in to it.

However, if you need the ground space or your compost is in a bin, then you can use a dark plastic bag or a garbage bin (with some holes drilled in it for drainage) to store your compost. If you're using a metal garbage can, you may want to spray paint with a non-toxic, rust-proof paint. You can then store the bags or bins in a dry area, such as a garage, shed or even a dry cellar. If storing in this way, you will need to keep an eye on the moisture content of your compost, stirring it when needed, to bring the damp compost on the bottom to the top; or, if your compost is dry all the way through, you'll need to mist and stir. You will also need to watch for mold, which can lead to you having to throw the entire bag or bin away.

Keep in mind that the longer that compost is stored, the more nutrients are used up and lost. To counteract this problem, some will try to add and mix in a little more browns and greens every so often, especially when storing the compost for more than a few months. Even if you think that the compost may have been stored a bit too long, don't throw it away—just mix it in with a new batch of compost that is nearing completion, or sprinkle it on as you layer new scraps on to your pile.

What about compost tea? Can it be stored?

Compost tea may also be stored, preferably for only a short time (again due to nutrient loss). The tea should be stored in a light-proof (dark) container with a good seal. Ideally, it should be stored for no more than six days. But, if you find that you need to store it for quite a bit longer, aeration is a must, as there are living organisms in your compost tea (an aquarium pump with a bubbler will work well for this).

IN CLOSING

. . . .

Composting is something that anyone can do, no matter where they live. Not only does it alleviate stress on our landfills, composting also provides us the nutrients needed to rejuvenate our soil, which can easily become exhausted when overused. Whether your garden consists of vegetables, fruits or even just a few flowers, the soil underneath will welcome a jolt of energy from the nutrient rich "black gold" that all gardeners prize.

Once you have completed your first compost cycle, you will almost certainly be hooked; if you haven't done so already, you'll soon have your second pile started, counting down the days until your next "soil harvest" (or "worm harvest," for vermiculturists).

While some individuals, towns and cities still need a bit of education—not only in regards to the benefits of compost, but the process as well—the ability to compost is worth fighting for. This is an opportunity for those of us who can appreciate the value of compost to shine, taking advantage of these teachable moments to show off the benefits and correct the misconceptions surrounding this beneficial resource.

So start saving that food garbage, and if your friends and family aren't composting (and you just can't get them interested), get their scraps as well. But beware: once they see what you end up with, you just might lose out on those extra scraps. You may even have visitors showing up, bearing bags to fill with your rich soil. Go ahead and share; you know where to get more!

Whether you decide to use bins, barrels, heaps or an indoor kitchen compost bucket, composting means you're doing everyone a favor: yourself, the earth, the landfills, and most of all your plants. And once your compost is finally ready . . .

Enjoy!

RESOURCES

Online

The Carbon Nitrogen Ratio
www.weblife.org/humanure/chapter3_7.html
A handy chart to assist with figuring your proper ratios.

Another Ratio Chart
www.plantnatural.com/composting-101/c-n-ratio/

Organic Calculator
www.organicsciencellc.com/composting.html

US Mortality and Butcher Waste Disposal Laws
http://compost.css.cornell.edu/mapsdisposal.html
A searchable map of the rules and regulations surrounding US
disposal laws as they relate to livestock composting.

Bins

Compost Bin Designs
www.compostguide.com/18-cool-diy-compost-bin-designs/

Tumbler Bin Plans
www.instructables.com/id/compost-bin/

www.motherearthnews.com/diy/compost-tumblers-zmaz79maz
raw.aspx

www.instructables.com/id/The-Best-Triple-Compost-Bin/

Choosing the Best Compost Bin
www.motherearthnews.com/organic-gardening/garden-tools-/
best-compost-bin-2m0z14fmzmar.aspx

Vermicomposting

www.rodalesorganiclife.com/gardenvermicomposting

Working Worms: How to Make Your Own Worm Farm
www.working-worms.com/how-to-make-your-own-worm-farm/

Books

Relf, D., & McDaniel, A. (2000). *Composting* (Rev. ed.). Blacksburg, Va.: Virginia Cooperative Extension.

Hirrel, S., & Riley, T. (1993). *Composting*. Fayetteville, Ark: Cooperative Extension Service, University of Arkansas, U.S. Dept. of Agriculture and county governments cooperating.

Composting livestock mortalities. (1997). Place of publication not identified: Ontario Ministry of Agriculture, Food and Rural Affairs.

Insam, H. (2002). *Microbiology of composting.* Berlin: Springer

Woolnough, M. (2010). *Worms & wormeries: Composting your kitchen waste-- and more!* Preston [England: Good Life Press.

Cullen, M., & Johnson, L. (1994). *The urban/suburban composter: The complete guide to backyard, balcony, and apartment composting.* New York: St. Martin's Press.

Backyard composting: Worm composting bin. (1995). Baton Rouge: Louisiana Cooperative Extension Service.

Cogger, C., & Sullivan, D. (2001). *Backyard composting* (Rev. Oct. 2001. ed.). Pullman, Wash.: Cooperative Extension, Washington State University.

Dickerson, G. (2000). *Vermicomposting.* Las Cruces, N.M.: Cooperative Extension Service, College of Agriculture and Home Economics, New Mexico State University.

Payne, B., & Bourgeois, P. (1999). *The worm cafe: Mid-scale vermicomposting of lunchroom wastes : A manual for schools, small businesses and community groups.* Kalamazoo, MI: Flower Press.

Vermicomposting (Rev: 08/92. ed.). (1992). Ontario Environment.

Ladner, P. (2011). *The urban food revolution: Changing the way we feed cities.* Gabriola Island, BC: New Society.

Cummings, D. (n.d.). *The organic composting handbook: Techniques for a healthy, abundant garden.*

Periodicals

Composting: An Urban Treasure.(Brief Article). (2001, September 22). *Earth Island Journal.*

Spagnoli, J. (1997, June 1). Waste not: Confessions of a backyard composter. *New York State Conservationist.*

Best, J. (2003, March 22). Compost this magazine! (Living Green). (composting techniques). *OnEarth.*

NOTES

NOTES

NOTES

NOTES

NOTES

NOTES

NOTES